ALGORITHMS
and DATA
STRUCTURES
in C++

CRC Press
Computer Engineering Series

Series Editor
Udo W. Pooch
Texas A&M University

Published books:

Telecommunications and Networking
Udo W. Pooch, Texas A&M University
Denis P. Machuel, Telecommunications Consultant
John T. McCahn, Networking Consultant

**Spicey Circuits: Elements of Computer-Aided
Circuit Analysis**
Rahul Chattergy, University of Hawaii

**Microprocessor-Based Parallel Architecture
for Reliable Digital Signal Processing Systems**
Alan D. George, Florida State University
Lois Wright Hawkes, Florida State University

Discrete Event Simulation: A Practical Approach
Udo W. Pooch, Texas A&M University
James A. Wall, Simulation Consultant

Algorithms and Data Structures in C++
Alan Parker, The University of Alabama in Huntsville

Forthcoming books:

Handbook of Software Engineering
Udo W. Pooch, Texas A&M University

ALGORITHMS
and DATA
STRUCTURES
in C++

Alan Parker
The University of Alabama in Huntsville

CRC Press
Boca Raton   Ann Arbor   London   Tokyo

Library of Congress Cataloging-in-Publication Data

Parker, Alan, 1959–
    Algorithms and data structures in C++ / Alan Parker.
        p. cm.—(CRC series in computer engineering)
    Includes bibliographical references and index.
    ISBN 0-8493-7171-6
    1. C++ (Computer program language). 2. Computer algorithms. 3. Data structures (Computer
science). I. Title. II. Series.
QA76.C153P37 1993
005.13′—dc20
DNLM/DLC
for Library of Congress   93-20537
                                                  CIP

No claim to original U.S. Government works
International Standard Book Number 0-8493-7171-6
Library of Congress Card Number 93-20537
Printed in the United States of America   2 3 4 5 6 7 8 9 0
Printed on acid-free paper

PREFACE

This text is designed for an introductory quarter or semester course in algorithms and data structures for students in engineering and computer science. It will also serve as a reference text for programmers in C++. The book presents algorithms and data structures with heavy emphasis on C++. Every C++ program presented is a stand-alone program. Except as noted, all of the programs in the book have been compiled and executed on multiple platforms.

When used in a course, the students should have access to C++ reference manuals for their particular programming environment. The instructor of the course should strive to describe to the students every line of each program. The prerequisite knowledge for this course should be a minimal understanding of digital logic. A high-level programming language is desirable but not required for more advanced students.

The study of algorithms is a massive field and no single text can do justice to every intricacy or application. The philosophy in this text is to choose an appropriate subset which exercises the unique and more modern aspects of the C++ programming language while providing a stimulating introduction to realistic problems.

I close with special thanks to my friend and colleague, Jeffrey H. Kulick, for his contributions to this manuscript.

Alan Parker
Huntsville, AL
1993

to

Valerie Anne Parker

LIST of FIGURES

LIST of PROGRAMS and OUTPUT

1 Data Representations

This chapter introduces the various formats used by computers for the representation of integers, floating point numbers, and characters. Extensive examples of these representations within the C++ programming language are provided.

1.1 Integer Representations

The tremendous growth in computers is partly due to the fact that physical devices can be built inexpensively which distinguish and manipulate two states at very high speeds. Since computers are devices which primarily act on two states (0 and 1), binary, octal, and hex representations are commonly used for the representation of computer data. The representation for each of these bases is shown in Table 1.1.

TABLE 1.1 Number Systems

Binary	Octal	Hexadecimal	Decimal
0	0	0	0
1	1	1	1
10	2	2	2
11	3	3	3
100	4	4	4
101	5	5	5
110	6	6	6
111	7	7	7
1000	10	8	8

TABLE 1.1 Number Systems (continued)

Binary	Octal	Hexadecimal	Decimal
1001	11	9	9
1010	12	A	10
1011	13	B	11
1100	14	C	12
1101	15	D	13
1110	16	E	14
1111	17	F	15
10000	20	10	16

Operations in each of these bases is analogous to base 10. In base 10, for example, the decimal number 743.57 is calculated as

$$743.57 = 7 \times 10^2 + 4 \times 10^1 + 3 \times 10^0 + 5 \times 10^{-1} + 7 \times 10^{-2} \qquad (1.1)$$

In a more precise form, if a number, X, has n digits in front of the decimal and m digits past the decimal

$$X = a_{n-1}a_{n-2}...a_1a_0.b_{m-1}b_{m-2}...b_1b_0 \qquad (1.2)$$

Its base 10 value would be

$$X = \sum_{j=0}^{n-1} a_j 10^j + \sum_{k=0}^{m-1} b_{m-1-k}10^{-k} \qquad 0 \le a_j, b_j \le 9 \qquad (1.3)$$

For hexadecimal,

$$X = \sum_{j=0}^{n-1} a_j 16^j + \sum_{k=0}^{m-1} b_{m-1-k}16^{-k} \qquad 0 \le a_j, b_j \le F \qquad (1.4)$$

For octal,

$$X = \sum_{j=0}^{n-1} a_j 8^j + \sum_{k=0}^{m-1} b_{m-1-k}8^{-k} \qquad 0 \le a_j, b_j \le 7 \qquad (1.5)$$

In general for base r

$$X = \sum_{j=0}^{n-1} a_j r^j + \sum_{k=0}^{m-1} b_{m-1-k} r^{-k} \qquad 0 \le a_j, b_j \le r-1 \qquad \text{(1.6)}$$

When using a theoretical representation to model an entity one can introduce a tremendous amount of bias into the thought process associated with the implementation of the entity. As an example, consider Eq. 1.6 which gives the value of a number in base r. In looking at Eq. 1.6, if a system to perform the calculation of the value is built, the natural approach is to subdivide the task into two subtasks: a subtask to calculate the integer portion and a subtask to calculate the fractional portion; however, this bias is introduced by the theoretical model. Consider, for instance, an equally valid model for the value of a number in base r. The number X is represented as

$$X = a_{n-1} a_{n-2} \ldots a_k . a_{k-1} \ldots a_0 \qquad \text{(1.7)}$$

where the decimal point appears after the kth element. X then has the value:

$$X = r^{-k} \left(\sum_{j=0}^{n-1} a_j r^j \right) \qquad \text{(1.8)}$$

Based on this model a different implementation might be chosen. While theoretical models are nice, they can often lead one astray.

As a first C++ programming example let's compute the representation of some numbers in decimal, octal, and hexadecimal for the integer type. A program demonstrating integer representations in decimal, octal, and hex is shown in Code List 1.1.

Code List 1.1 Integer Example

C++ Source Program
#include <iostream.h>
int a[]={45,245,567,1014,–45,–1,256};
void main()
{
int i;
for(i=0;i<sizeof(a)/sizeof(int);i++)
{
cout << endl << endl << "In decimal " << dec << a[i];

Code List 1.1 Integer Example (continued)

C++ Source Program
cout << endl << "In hex " << hex << a[i];
cout << endl << "In octal " << oct << a[i];
}
}

In this sample program there are a couple of C++ constructs. The *#include <iostream.h>* includes the header files which allow the use of *cout*, a function used for output. The second line of the program declares an array of integers. Since the list is initialized the size need not be provided. This declaration is equivalent to

int a[7]; — declaring an array of seven integers 0–6
a[0]=45; — initializing each entry
a[1]=245;
a[2]=567;
a[3]=1014;
a[4]=–45;
a[5]=–1;
a[6]=256;

The *void main()* declaration declares that the main program will not return a value. The *sizeof* operator used in the loop for *i* returns the size of the array *a* in bytes. For this case

sizeof(a)=28

sizeof(int)=4

The *cout* statement in C++ is used to output the data. It is analogous to the *printf* statement in C but without some of the overhead. The *dec*, *hex*, and *oct* keywords in the *cout* statement set the output to decimal, hexadecimal, and octal respectively. The default for *cout* is in decimal.

At this point, the output of the program should not be surprising except for the representation of negative numbers. The computer uses a 2's complement representation for numbers which is discussed in Section 1.1.3 on page 7.

Code List 1.2 Program Output of Code List 1.1

C++ Output
In decimal 45
In hex 2d
In octal 55
In decimal 245
In hex f5
In octal 365
In decimal 567
In hex 237
In octal 1067
In decimal 1014
In hex 3f6
In octal 1766
In decimal –45
In hex ffffffd3
In octal 37777777723
In decimal –1
In hex ffffffff
In octal 37777777777
In decimal 256
In hex 100
In octal 400

1.1.1 Unsigned Notation

Unsigned notation is used to represent nonnegative integers. The unsigned notation does not support negative numbers or floating point numbers. An n-bit number, A, in unsigned notation is represented as

$$A \equiv a_{n-1}a_{n-2}\cdots a_0 \tag{1.9}$$

with a value of

$$A = \sum_{k=0}^{n-1} a_k 2^k \qquad a_k \in \{0, 1\} \tag{1.10}$$

Negative numbers are not representable in unsigned format. The range of numbers in an n-bit unsigned notation is

$$0 \le A \le 2^n - 1 \tag{1.11}$$

Zero is uniquely represented in unsigned notation. The following types are used in the C++ programming language to indicate unsigned notation:

- unsigned char (8 bits)
- unsigned short (16 bits)
- unsigned int (native machine size)
- unsigned long (machine dependent)

The number of bits for each type can be compiler dependent.

1.1.2 Signed-Magnitude Notation

Signed-magnitude numbers are used to represent positive and negative integers. Signed-magnitude notation does not support floating-point numbers. An n-bit number, A, in signed-magnitude notation is represented as

$$A \equiv a_{n-1}a_{n-2}\cdots a_0 \tag{1.12}$$

with a value of

$$A = (-1)^{a_{n-1}} \left(\sum_{k=0}^{n-2} a_k 2^k \right) \qquad a_k \in \{0, 1\} \tag{1.13}$$

A number, A, is negative if and only if $a_{n-1} = 1$. The range of numbers in an n-bit signed magnitude notation is

$$-(2^{n-1} - 1) \le A \le 2^{n-1} - 1 \tag{1.14}$$

The range is symmetrical and zero is not uniquely represented. Computers do not use signed-magnitude notation for integers because of the hardware complexity induced by the representation to support addition.

1.1.3 2's Complement Notation

2's complement notation is used by almost all computers to represent positive and negative integers. An n-bit number, A, in 2's complement notation is represented as

$$A \equiv a_{n-1}a_{n-2}...a_0 \tag{1.15}$$

with a value of

$$A = \left(\sum_{k=0}^{n-2} a_k 2^k \right) - a_{n-1} 2^{n-1} \qquad a_k \in \{0, 1\} \tag{1.16}$$

A number, A, is negative if and only if $a_{n-1} = 1$. From Eq. 1.16, the negative of A, $-A$, is given as

$$-A = \left(\sum_{k=0}^{n-2} -a_k 2^k \right) + a_{n-1} 2^{n-1} \tag{1.17}$$

which can be written as

$$-A = 1 + \left(\sum_{k=0}^{n-2} (\overline{a_k}) 2^k \right) - \overline{a_{n-1}} 2^{n-1} \tag{1.18}$$

where $\bar{x}$ is defined as the unary complement:

$$\bar{x} = \begin{cases} 1, & \text{if } x = 0 \\ 0, & \text{if } x = 1 \end{cases} \tag{1.19}$$

The one's complement of a number, A, denoted by $\overline{A}$, is defined as

$$\overline{A} = \overline{a_{n-1}} \, \overline{a_{n-2}} ... \overline{a_0} \tag{1.20}$$

From Eq. 1.18 it can be shown that

$$-A = 1 + \overline{A} \tag{1.21}$$

To see this note that

$$\bar{A} = -\overline{a_{n-1}}2^{n-1} + \sum_{k=0}^{n-2} \overline{a_k}2^k \qquad \text{(1.22)}$$

and

$$\sum_{k=0}^{n-2} \overline{a_k}2^k + \sum_{k=0}^{n-2} a_k 2^k$$

$$= \sum_{k=0}^{n-2} (\overline{a_k} + a_k)\, 2^k = \sum_{k=0}^{n-2} 2^k = 2^{n-1} - 1 \qquad \text{(1.23)}$$

This yields

$$\sum_{k=0}^{n-2} \overline{a_k}2^k = 2^{n-1} - 1 - \sum_{k=0}^{n-2} a_k 2^k \qquad \text{(1.24)}$$

Inserting Eq. 1.24 into Eq. 1.22 yields

$$\bar{A} + 1 = -\overline{a_{n-1}}2^{n-1} + 2^{n-1} - 1 - \sum_{k=0}^{n-2} a_k 2^k + 1 \qquad \text{(1.25)}$$

which gives

$$\bar{A} + 1 = (1 - \overline{a_{n-1}})\, 2^{n-1} - \sum_{k=0}^{n-2} a_k 2^k \qquad \text{(1.26)}$$

By noting

$$1 - \overline{a_{n-1}} = a_{n-1} \qquad \text{(1.27)}$$

one obtains

$$\bar{A} + 1 = a_{n-1}2^{n-1} - \sum_{k=0}^{n-2} a_k 2^k \qquad \text{(1.28)}$$

which is $-A$. So whether A is positive or negative the two's complement of A is equivalent to $-A$.

$$
\begin{array}{l}
0000\ 0001 = +1 \\
1111\ 1110\ \text{(8-bit 1's complement)} \\
\underline{\qquad +1} \\
1111\ 1111 = -1\ \text{(8-bit 2's complement)}
\end{array}
$$

Note that in this case it is a simpler way to generate the representation of -1. Otherwise you would have to note that

$$- 128 + 64 + 32 + 16 + 8 + 4 + 2 + 1 = -1 \qquad (1.29)$$

Similarly

$$1111\ 1111 = -1$$
$$0000\ 0000\ \text{(8-bit 1's complement)}$$
$$\underline{+1}$$
$$0000\ 0001 = 1\ \text{(8-bit 2's complement)}$$

However, it is useful to know the representation in terms of the weighted bits. For instance, -5, can be generated from the representation of -1 by eliminating the contribution of 4 in -1:

weight of 4

$$-1 = 1111\ 1111 \qquad \textbf{\underline{8-bit 2's complement}}$$
$$-5 = 1111\ 1011$$

Similarly, -21, can be realized from -5 by eliminating the positive contribution of 16 from its representation.

weight of 16

$$-5 = 1111\ 1011 \qquad \textbf{\underline{8-bit 2's complement}}$$
$$-21 = 1110\ 1011$$

The operations can be done in hex as well as binary. For 8-bit 2's complement one has

$$-1 = FF \qquad (1.30)$$

$$1 = \overline{FF} + 1 = 00 + 1 = 01 \qquad (1.31)$$

with all the operations performed in hex. After a little familiarity, hex numbers are generally easier to manipulate. To take the one's complement one handles each hex digit at a time. If w is a hex digit then the 1's complement of w, $\overline{w}$, is given as

$$\overline{w} = F - w \qquad (1.32)$$

$$\overline{A} = F - A = 5 \qquad (1.33)$$

The range of numbers in an n-bit 2's complement notation is

$$-2^{n-1} \leq A \leq 2^{n-1} - 1 \tag{1.34}$$

The range is not symmetric but the number zero is uniquely represented.

The representation in 2's complement arithmetic is similar to an odometer in a car. If the car odometer is reading zero and the car is driven one mile in reverse (-1) then the odometer reads 999999. This is illustrated in Table 1.2.

TABLE 1.2 2's Complement Odometer Analogy

8-Bit 2's Complement		
Binary	Value	Odometer
11111110	-2	999998
11111111	-1	999999
00000000	0	000000
00000001	1	000001
00000010	2	000002

Typically, 2's complement representations are used in the C++ programming language with the following declarations:

- char (8 bits)
- short (16 bits)
- int (16,32, or 64 bits)
- long (32 bits)

The number of bits for each type can be compiler dependent. An 8-bit example of the three basic integer representations is shown in Table 1.3.

TABLE 1.3 8-Bit Representations

8-Bit Representations			
Number	Unsigned	Signed Magnitude	2's Complement
-128	NR†	NR	10000000

TABLE 1.3 8-Bit Representations (continued)

8-Bit Representations			
Number	Unsigned	Signed Magnitude	2's Complement
-127	NR	11111111	10000001
-2	NR	10000010	11111110
-1	NR	10000001	11111111
0	00000000	00000000 10000000	00000000
1	00000001	00000001	00000001
127	01111111	01111111	01111111
128	10000000	NR	NR
255	11111111	NR	NR

†. Not representable in 8-bit format.

TABLE 1.4 Ranges for 2's Complement and Unsigned Notations

# Bits	2's Complement	Unsigned
8	$-128 \leq A \leq 127$	$0 \leq A \leq 255$
16	$-32768 \leq A \leq 32767$	$0 \leq A \leq 65535$
32	$-2147483648 \leq A \leq 2147483647$	$0 \leq A \leq 4294967295$
n	$-2^{n-1} \leq A \leq 2^{n-1} - 1$	$0 \leq A \leq 2^n - 1$

The ranges for 8-, 16-, and 32-bit representations for 2's complement and unsigned representations are shown in Table 1.4.

1.1.4 Sign Extension

This section investigates the conversion from an n-bit number to an m-bit number for signed-magnitude, unsigned, and 2's complement. It is assumed that $m>n$. This problem is important due to the fact that many processors use different sizes for their operands. As a result, to move data from one processor to another requires a conversion. A typical problem might be to convert 32-bit formats to 64-bit formats.

Given A as

$$A \equiv a_{n-1}a_{n-2}\ldots a_0 \tag{1.35}$$

and B as

$$B \equiv b_{m-1}b_{m-2}\ldots b_0 \tag{1.36}$$

the objective is to determine b_k such that $B = A$.

1.1.4.1 Signed-Magnitude

For signed-magnitude the b_k are assigned with

$$b_k = \begin{cases} a_k, & 0 \le k \le n-2 \\ 0, & n \le k \le m-2 \\ a_{n-1}, & k = m-1 \end{cases} \tag{1.37}$$

1.1.4.2 Unsigned

The conversion for unsigned results in

$$b_k = \begin{cases} a_k, & 0 \le k \le n-1 \\ 0, & n \le k < m \end{cases} \tag{1.38}$$

1.1.4.3 2's Complement

For 2's complement there are two cases depending on the sign of the number:

(a) $(a_{n-1} = 0)$ For this case, A reduces to

$$A = \sum_{k=0}^{n-2} a_k 2^k \tag{1.39}$$

It is trivial to see that the assignment of b_k with

$$b_k = \begin{cases} a_k, & 0 \le k \le n-2 \\ 0, & n-1 \le k < m \end{cases} \qquad (1.40)$$

satisfies this case.

(b) ($a_{n-1} = 1$) For this case

$$A = \left(\sum_{k=0}^{n-2} a_k 2^K \right) - 2^{n-1} \qquad (1.41)$$

By noting that

$$\left(\sum_{k=n-1}^{m-2} 2^k \right) - 2^{m-1} = -2^{n-1} \qquad (1.42)$$

The assignment of b_k with

$$b_k = \begin{cases} a_k, & 0 \le k \le n-2 \\ 1, & n-1 \le k < m \end{cases} \qquad (1.43)$$

satisfies the condition. The two cases can be combined into one assignment with b_k as

$$b_k = \begin{cases} a_k, & 0 \le k \le n-2 \\ a_{n-1}, & n-1 \le k < m \end{cases} \qquad (1.44)$$

The sign, a_{n-1}, of A is simply extended into the higher order bits of B. This is known as sign-extension. Sign extension is illustrated from 8-bit 2's complement to 32-bit 2's complement in Table 1.5.

TABLE 1.5 2's Complement Sign Extension

8-Bit	32-Bit
0xff	0xffffffff
0x0f	0x0000000f
0x01	0x00000001
0x80	0xffffff80
0xb0	0xffffffb0

1.1.5 C++ Program Example

This section demonstrates the handling of 16-bit and 32-bit data by two different processors. A simple C++ source program is shown in Code List 1.3. The assembly code generated for the C++ program is demonstrated for the Intel 80286 and the Motorola 68030 in Code List 1.4. A line-by-line description follows:

- Line # 1: The 68030 executes a *movew* instruction moving the constant 1 to the address where the variable *i* is stored. The *movew*—move word—instruction indicates the operation is 16 bits.

  The 80286 executes a *mov* instruction. The *mov* instruction is used for 16-bit operations.
- Line # 2: Same as Line # 1 with different constants being moved.
- Line # 3: The 68030 moves *j* into register *d0* with the *movew* instruction. The *addw* instruction performs a word (16-bit) addition storing the result at the address of the variable *i*.

  The 80286 executes an *add* instruction storing the result at the address of the variable *i*. The instruction does not involve the variable *j*. The compiler uses the immediate data, 2, since the assignment of *j* to 2 was made on the previous instruction. This is a good example of optimization performed by a compiler. An unoptimizing compiler would execute

  mov ax, WORD PTR [bp-4]
  add WORD PTR [bp-2], ax

  similar to the 68030 example.
- Line # 4: The 68030 executes a *moveq*—quick move—of the immediate data 3 to register *d0*. A long move, *movel*, is performed moving the value to the address of the variable *k*. The long move performs a 32-bit move.

  The 80286 executes two immediate moves. The 32-bit data is moved to the address of the variable k in two steps. Each step consists of a 16-bit move. The least significant word, 3, is moved first followed by the most significant word,0.
- Line # 5: Same as Line # 4 with different constants being moved.
- Line # 6: The 68030 performs an add long instruction, *addl*, placing the result at the address of the variable *k*.

  The 80286 performs the 32-bit operation in two 16-bit instructions. The first part consists of an add instruction, *add*, followed by an add with carry instruction, *adc*.

Code List 1.3 Assembly Language Example

Line #	C Code
	void main()
	{
	short i,j;
	long k,l;
1	i=1;
2	j=2;
3	i=i+j;
4	k=3;
5	l=4;
6	k=k+l;
	}

Code List 1.4 Assembly Language Code

Line#	68030	80286
1	movew #1,a6@(−2)	mov WORD PTR [bp-2],1
2	movew #2,a6@(−4)	mov WORD PTR [bp-4],2
3	movew a6@(−4),d0	add WORD PTR [bp-2],2
	addw d0,a6@(−2)	
4	moveq #3,d0	mov WORD PTR [bp-8],3
	movel d0,a6@(−8)	mov WORD PTR [bp-6],0
5	moveq #4,d0	mov WORD PTR [bp-12],4
	movel d0,a6@(−12)	mov WORD PTR [bp-10],0
6	addl d0,a6@(−8)	add WORD PTR [bp-8],4
		adc WORD PTR [bp-6],0

This example demonstrates that each processor handles different data types with different instructions. This is one of the reasons that the high level language requires the declaration of specific types.

1.2 Floating Point Representation

1.2.1 IEEE 754 Standard Floating Point Representations

Floating point is the computer's binary equivalent of scientific notation. A floating point number has both a fraction value or mantissa and an exponent value. In high level languages floating point is used for calculations involving real numbers. Floating point operation is desirable because it eliminates the need for careful problem scaling. IEEE Standard 754 binary floating point has become the most widely used standard. The standard specifies a 32-bit, a 64-bit, and an 80-bit format.

1.2.1.1 IEEE 32-Bit Standard

The IEEE 32-bit standard is often referred to as single precision format. It consists of a 23-bit fraction or mantissa, f, an 8-bit biased exponent, e, and a sign bit, s. Results are normalized after each operation. This means that the most significant bit of the fraction is forced to be a one by adjusting the exponent. Since this bit must be one it is not stored as part of the number. This is called the implicit bit. A number then becomes

$$(-1)^s (1.f) 2^{e-127} \tag{1.45}$$

The number zero, however, cannot be scaled to begin with a one. For this case the standard indicates that 32-bits of zeros is used to represent the number zero.

1.2.1.2 IEEE 64-bit Standard

The IEEE 64-bit standard is often referred to as double precision format. It consists of a 52-bit fraction or mantissa, f, an 11-bit biased exponent, e, and a sign bit, s. As in single precision format the results are normalized after each operation. A number then becomes

$$(-1)^s (1.f) 2^{e-1023} \tag{1.46}$$

The number zero, however, cannot be scaled to begin with a one. For this case the standard indicates that 64-bits of zeros is used to represent the number zero.

1.2.1.3 C++ Example for IEEE Floating point

A C++ source program which demonstrates the IEEE floating point format is shown in Code List 1.5.

Code List 1.5 C++ Source Program

C++ Source
| ```
#include <stdio.h>
#include <iostream.h>
#include <iomanip.h>
union float_point_32 {
        float fp;
        long li;
        float_point_32(float in= float(0.0))
                {
                fp=in; }
        };


union float_point_64 {
        double fp;
        long li[2];
        float_point_64(double in = double(0.0))
                {
                fp=in; }
 };
class float_number_32 {
        float_point_32 f;
        public:
        float_number_32(float in=float(0.0)) {f.fp=in;}
        float fp() { return f.fp;}
        long li() { return f.li;}
        long sign() {return f.li&0x80000000?1:0; }
        long exponent()
                { return (f.li&0x7f800000)>>23;}
        void fraction();
        void print() {
        cout << "Floating point = " << f.fp <<
        " 32-bit Representation = " << hex << setfill('0') << setw(8)
        << f.li << dec << endl
``` |
```

Code List 1.5   C++ Source Program (continued)

C++ Source

```cpp
                    << "sign = " << sign() << " biased exponent = " << exponent() <<
                    " unbiased exponent = " << exponent()–127
                    << endl << "fraction = ";
                    fraction();
                    cout << endl << endl;}
                    };
void float_number_32::fraction()
            {
            unsigned long mask=0x400000;
            int i;
             if (f.li==0) cout << "0."; else cout << "1.";
            for(i=0;i<23;i++) {
                            if(mask&f.li) cout << "1";
                            else cout << "0";
                            mask = mask >> 1; }
            }
class float_number_64 {
            float_point_64 f;
            public:
            float_number_64(double in=double(0.0)) {f.fp=in;}
            double fp() { return f.fp;}
            long li_ms() { return f.li[1];}
            long li_ls() { return f.li[0];}
            long sign() {return f.li[0]&0x80000000?1:0; }
            long exponent()
                        { return (f.li[1]&0x7ff00000)>>20;}
            void fraction();
            void print() {
            cout << "Floating point = " << f.fp <<
            " 64-bit Representation = " << hex << setfill('0') << setw(8)
            << li_ms() << setw(8) << li_ls() << dec << endl;
            << "sign = " << sign() << " biased exponent = " << exponent() <<
```

Code List 1.5 C++ Source Program (continued)

C++ Source

```
                " unbiased exponent = " << exponent()–1023
                << endl << "fraction = ";
                fraction();
                cout << endl << endl;}
                };
void float_number_64::fraction()
                {
                unsigned long mask=0x80000;
                int i;
                if ((f.li[0]||f.li[1])==0) cout << "0."; else
                        cout << "1.";
                        for(i=0;i<20;i++) {
                        if(mask&f.li[1]) cout << "1";
                        else cout << "0";
                        mask = mask >> 1; }
                        mask=0x80000000;
                        for(i=0;i<32;i++) {
                        if(mask&f.li[0]) cout << "1";
                        else cout << "0";
                        mask = mask >> 1; }
                }
void main()
{
                float_number_32 x(0.1),y(–5.0);
                float_number_64 z(0.1);
                x.print();
                y.print();
                z.print();
}
```

The output of the program is shown in Code List 1.6. The *union* operator allows a specific memory location to be treated with different types. For this case the memory location holds 32 bits. It can be treated as a *long* integer (an integer of 32 bits) or a floating point number. The *union* operator is necessary for this program because bit operators in C and C++ do not operate on floating point numbers. The *float_point_32(float in=float(0.0)) {fp=in}* function demonstrates the use of a constructor in C++. When a variable is declared to be of type *float_point_32* this function called. If a parameter is not specified in the declaration then the default value, for this case 0.0, is assigned. A declaration of *float_point_32 x(0.1),y;* therefore, would initialize x.fp to 0.1 and y.fp to 0.0.

Code List 1.6   Output of Program in Code List 1.5

C++ Output
Floating point = 0.1 32-bit representation = 3dcccccd
sign = 0 biased exponent = 123 unbiased exponent = –4
fraction = 1.10011001100110011001101
Floating point = –5 32-bit representation = c0a00000
sign = 1 biased exponent = 129 unbiased exponent = 2
fraction = 1.01000000000000000000000
Floating point = 0.1 64-bit representation = 3fb999999999999a
sign = 1 biased exponent = 1019 unbiased exponent = –4
fraction = 1.1001100110011001100110011001100110011001100110011010

The *union float_point_64* declaration allows 64 bits in memory to be thought of as one 64-bit floating point number(double) or 2 32-bit long integers. The *void float_number_32::fraction()* demonstrates scoping in C++. For this case the function *fraction()* is associated with the class *float_number_32*. Since *fraction* was declared in the public section of the class *float_-number_32* the function has access to all of the public and private functions and data associated with the class *float_number_32*. These functions and data need not be declared in the function. Notice for this example *f.li* is used in the function and only mask and *i* are declared locally. The *setw()* used in the *cout* call in *float_number_64* sets the precision of the output. The program uses a number of bit operators in C++ which are described in the next section.

1.2.2   Bit Operators in C++

C++ has bitwise operators &, ^, |, and ~. The operators &, ^, and | are binary operators while the operator ~ is a unary operator.

- ~, 1's complement
- &, bitwise and
- ^, bitwise exclusive or
- |, bitwise or

The behavior of each operator is shown in Table 1.6.

TABLE 1.6 Bit Operators in C++

a	b		a&b	a^b	a|b	~a
0	0		0	0	0	1
0	1		0	1	1	1
1	0		0	1	1	0
1	1		1	0	1	0

To test out the derivation for calculating the 2's complement of a number derived in Section 1.1.3 a program to calculate the negative of a number is shown in Code List 1.7. The output of the program is shown in Code List 1.8. Problem 1.11 investigates the output of the program.

Code List 1.7 Testing the Binary Operators in C++

```
C++ Source Code

#include <iostream.h>
class data
{
int x;
public:
        void set(int y) { x=y;}
        void print() {
        cout << "The value of x is " << x << endl;
        cout << "The value of the 2's complement of x is "
                << ~x+1 << endl << endl;
        }
};
void main()
{
data x;
x.set(7); x.print();
```

Code List 1.7 Testing the Binary Operators in C++ (continued)

C++ Source Code
x.set(–100); x.print();
x.set(0); x.print();
x.set(1<<sizeof(int)*8–1); x.print();
x.set((1<<sizeof(int)*8–1)+1); x.print();
}

Code List 1.8 Output of Program in Code List 1.7

C++ Output
The value of x is 7
The value of the 2's complement of x is –7
The value of x is –100
The value of the 2's complement of x is 100
The value of x is 0
The value of the 2's complement of x is 0
The value of x is –32768
The value of the 2's complement of x is –32768
The value of x is –32767
The value of the 2's complement of x is 32767

A program demonstrating one of the most important uses of the OR operator, |, is shown in Code List 1.9. The output of the program is shown in Code List 1.10. Figure 1.1 demonstrates the value of x for the program. The eight attributes are packed into one character. The character field can hold $256 = 2^8$ combinations handling all combinations of each attribute taking on the value ON or OFF. This is the most common use of the OR operators. For a more detailed example consider the file operation command for opening a file. The file definitions are defined in <iostream.h> by BORLAND C++ as shown in Table 1.7.

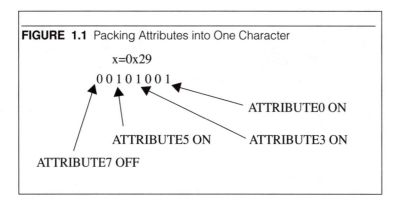

FIGURE 1.1 Packing Attributes into One Character

Code List 1.9 Bit Operators

C++ Source Code
#include <iostream.h>
#define ATTRIBUTE0 0x1
#define ATTRIBUTE1 0x2
#define ATTRIBUTE2 0x4
#define ATTRIBUTE3 0x8
#define ATTRIBUTE4 0x10
#define ATTRIBUTE5 0x20
#define ATTRIBUTE6 0x40
#define ATTRIBUTE7 0x80
typedef unsigned char attribute;
void main()
{
attribute x;
x=ATTRIBUTE0|ATTRIBUTE3|ATTRIBUTE5;
//Test to see if a has desired attribute;
if(x&ATTRIBUTE6) cout << "x has attribute ATTRIBUTE6" << endl;
else cout << "x does not have attribute ATTRIBUTE6" << endl;
if(x&ATTRIBUTE3) cout << "x has attribute ATTRIBUTE3" << endl;
else cout << "x does not have attribute ATTRIBUTE3" << endl;
if(x&(ATTRIBUTE2|ATTRIBUTE0)) cout << "x has at least one of the attributes:"

Code List 1.9   Bit Operators (continued)

C++ Source Code
" ATTRIBUTE2, ATTRIBUTE0"<< endl; cout << "x has a hex value: " << hex <<(int) x << endl; }

Code List 1.10   Output of Program in Code List 1.9

C++ Output
x does not have attribute ATTRIBUTE6
x has attribute ATTRIBUTE3
x has at least one of the attributes: ATTRIBUTE2, ATTRIBUTE0
x has a hex value: 29

TABLE 1.7   Fields for File Operations in C++

Source
enum open_mode {
in = 0x01, // open for reading
out = 0x02, // open for writing
ate = 0x04, // seek to eof upon original open
app = 0x08, // append mode: all additions at eof
trunc = 0x10, // truncate file if already exists
nocreate = 0x20, // open fails if file doesn't exist
noreplace= 0x40, // open fails if file already exists
binary = 0x80 // binary (not text) file
};

A program illustrating another use is shown in Code List 1.11. If the program executes correctly the output file, test.dat, is created with the string, "This is a test", placed in it. The file, test.dat, is opened for writing with *ios::out* and for truncation with *ios::trunc*. The two modes are presented together to the *ofstream* constructor with the use of the *or* function.

Code List 1.11   Simple File I/O

C++ Source
| ```
#include <fstream.h>
void main()
        {
        ofstream file("test.dat",ios::outlios::trunc);
        if(!file)
                {
                cout << "Could not open file"<< endl;
                return;
                }
        file << "This is a test";
        }
``` |

1.2.3 Examples

This section presents examples of IEEE 32-bit and 64-bit floating point representations. Converting 100.5 to IEEE 32-bit notation is demonstrated in Example 1.1.

Determining the value of an IEEE 64-bit number is shown in Example 1.2. In many cases for problems as in Example 1.1 the difficulty lies in the actual conversion from decimal to binary. The next section presents a simple methodology for such a conversion.

1.2.4 Conversion from Decimal to Binary

This section presents a simple methodology to convert a decimal number, A, to its corresponding binary representation. For the sake of simplicity, it is assumed the number satisfies

$$0 \le A < 1 \tag{1.47}$$

in which case we are seeking the a_k such that

$$A = \sum_{k=1}^{\infty} a_k 2^{-k} \tag{1.48}$$
```

EXAMPLE 1.1   IEEE 32-Bit Format

Representing 100.5 in IEEE 32-Bit Format

Step 1 Convert number to binary
$$100.5 = 1100100.1$$

Step 2 Scale the number so the fraction
begins with a 1
$$100.5 = 1.1001001 \times 2^6$$

Step 3 For sign bit place 0 if positive, 1 if negative
sign bit = 0

Step 4 Calculate 8-bit exponent field in binary
$$exp = 127 + 6 = 133 = 10000101$$

Step 5 Strip fraction which follows 1.
fraction = 1001001

Step 6 Combine bits together
0 10000101 10010010000000000000000
s   exp   fraction

Step 7 Convert to hex
42C90000

Answer: 42C90000

The simple procedure is illustrated in Code List 1.12. The C Code performing the decimal to binary conversion is shown in Code List 1.13. The output of the program is shown in Code List 1.14. This program illustrates the use of the default value. When a variable is declared as z is by *data z*, z is assigned 0.0 and *precision* is assigned 32. This can be seen as in the program *z.prec()* is never called and the output results in 32 bits of precision. The paper conversion for 0.4 is illustrated in Example 1.3.

1.3   Character Formats—ASCII

To represent keyboard characters, a standard has been adopted to ensure compatibility across many different machines. The most widely used standard is

---

EXAMPLE 1.2   Calculating the Value of an IEEE 64-Bit
            Number

Find the value of the IEEE 64-bit number given by the 16 hex digits

4042900000000000

Step 1 Convert to binary and identify sign, exponent, and fraction.

0100 0000 0100 0010 1001 0000... 0000

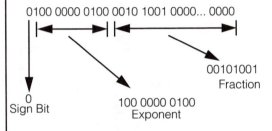

00101001
Fraction

0
Sign Bit

100 0000 0100
Exponent

Step 2 Calculate Exponent

Exponent = 100 0000 0100 = 1028
Exponent value = 1028–1023 = 5

Step 3 Calculate Fraction
Add implicit 1
Fraction = 1.00101001

Step 4 Put it together

Result = 1.00101001×2^5

= 100101.001 = 37.125

Answer: 37.125

---

the ASCII (American Standard Code for Information Interchange) character set. This set has a one byte format and is shown in Table 1.8. It allows for 256 distinct characters and specifies the first 128. The lower ASCII characters are control characters which were derived from their common use in earlier machines.Although the ASCII standard is widely used, different operating systems use different file formats to represent data, even when the data files contain only characters. Two of the most popular systems, DOS and Unix differ in their file format. For example, the text file shown in Table 1.9 has a DOS format shown in Table 1.10 and a Unix format shown in Table 1.11. Notice that the DOS file use a carriage return, cr, followed by a new line, nl, while the Unix file uses only a new line. As a result Unix text files will be smaller than

EXAMPLE 1.3   Converting 0.4 from Decimal to Binary

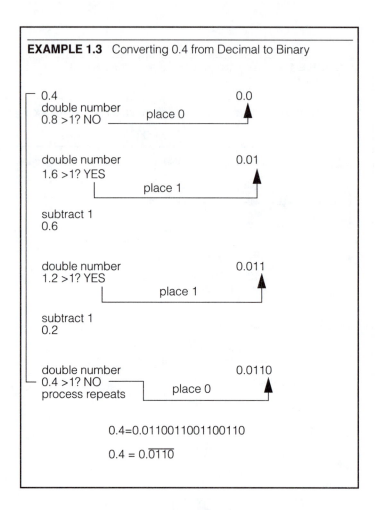

0.4
double number
0.8 >1? NO   place 0   0.0

double number   0.01
1.6 >1? YES
       place 1

subtract 1
0.6

double number   0.011
1.2 >1? YES
       place 1

subtract 1
0.2

double number   0.0110
0.4 >1? NO
process repeats   place 0

0.4=0.0110011001100110

$0.4 = 0.0\overline{0110}$

Code List 1.12   Decimal to Binary Conversion

Pseudo-Code

```
k=1
While (More Precision Required)
        {
        A ← 2A
        if A>1
                {
                A=A–1
                a_k = 1
                }
        else a_k = 0
        k=k+1
        }
```

Code List 1.13 Decimal to Conversion C++ Program

C++ Source

```cpp
// This program demonstrates the conversion of a decimal number
// to a binary number for numbers of type double
// Numbers must be of the form 0.xxxxxxxxxxx
#include <iostream.h>
class data {
        double d;
 unsigned int precision;
        public:
        data(double in=0.0) { d=in;precision=32;}
        void prec(unsigned int p) { precision=p;}
        void binary_calc(double in);
        void value() { cout << "Decimal value = " << d << endl; }
        void binary() { cout << "Binary value = "; binary_calc(d);
        cout << endl << endl;}
        };
void data::binary_calc(double in)
        {
        int i;
         cout << "0." ; // program works on this format of numbers only
        for(i=0;i<precision;i++)
                {
                in*=2.0;
                if(in>=1)
                        {
                        in-=1;
                        cout << "1";
                        }
                else cout << "0";
                };
        }
void main()
        {
```

Code List 1.13   Decimal to Conversion C++ Program (continued)

C++ Source

data x(0.7), y(0.1), z;
 	
x.prec(20);
 	
x.value();
 	
x.binary();
 	
y.prec(32);
 	
y.value();
 	
y.binary();
 	
z.value();
 	
z.binary();
 	
}

Code List 1.14   Output of Program in Code List 1.13

C++ Output
Decimal value = 0.7 Binary value = 0.10110011001100110011 Decimal value = 0.1 Binary value = 0.00011001100110011001100110011001 Decimal value = 0 Binary value = 0.00000000000000000000000000000000

DOS text files. In the DOS and Unix tables, underneath each character is its ASCII representation in hex. The numbering on the left of each table is the offset in octal of the line in the file.

TABLE 1.8   ASCII Listing

ASCII Listing							
00 nul	01 soh	02 stx	03 etx	04 eot	05 enq	06 ack	07 bel
08 bs	09 ht	0a nl	0b vt	0c np	0d cr	0e so	0f si
10 dle	11 dc1	12 dc2	13 dc3	14 dc4	15 nak	16 syn	17 etb
18 can	19 em	1a sub	1b esc	1c fs	1d gs	1e rs	1f us

TABLE 1.8 ASCII Listing (continued)

ASCII Listing							
20 sp	21 !	22 "	23 #	24 $	25 %	26 &	27 '
28 (	29)	2a *	2b +	2c ,	2d -	2e .	2f /
30 0	31 1	32 2	33 3	34 4	35 5	36 6	37 7
38 8	39 9	3a :	3b ;	3c <	3d =	3e >	3f ?
40 @	41 A	42 B	43 C	44 D	45 E	46 F	47 G
48 H	49 I	4a J	4b K	4c L	4d M	4e N	4f O
50 P	51 Q	52 R	53 S	54 T	55 U	56 V	57 W
58 X	59 Y	5a Z	5b [	5c \	5d]	5e ^	5f _
60	61 a	62 b	63 c	64 d	65 e	66 f	67 g
68 h	69 i	6a j	6b k	6c l	6d m	6e n	6f o
70 p	71 q	72 r	73 s	74 t	75 u	76 v	77 w
78 x	79 y	7a z	7b {	7c l	7d }	7e ~	7f del

TABLE 1.9 Text File

Test File
This is a test file
We will look at this file under Unix and DOS

1.4 Putting it All Together

This section presents an example combining ASCII, floating point, and integer types using one final C++ program. The program is shown in Code List 1.15 and the output is shown in Code List 1.16.

The program utilizes a common memory location to store 8 bytes of data. The data will be treated as double, float, char, int, or long. A particular memory implementation for this program is shown in Figure 1.2.

TABLE 1.10 DOS File Format

DOS File Format																
0000000	T	h	i	s	sp	i	s	sp	a	sp	t	e	s	t	sp	f
	5468		6973		2069		7320		6120		7465		7374		2066	
0000020	i	l	e	cr	nl	W	e	sp	w	i	l	l	sp	l	o	o
	696c		650d		0a57		6520		7769		6c6c		206c		6f6f	
0000040	k	sp	a	t	sp	t	h	i	s	sp	f	i	l	e	sp	u
	6b20		6174		2074		6869		7320		6669		6c65		2075	
0000060	n	d	e	r	sp	U	n	i	x	sp	a	n	d	sp	D	O
	6e64		6572		2055		6e69		7820		616e		6420		444f	
0000100	S	cr	nl													
	530d		0a00													
0000103																

TABLE 1.11 Unix File Format (ISO)

ISO File Format																
0000000	T	h	i	s	sp	i	s	sp	a	sp	t	e	s	t	sp	f
	5468		6973		2069		7320		6120		7465		7374		2066	
0000020	i	l	e	nl	W	e	sp	w	i	l	l	sp	l	o	o	k
	696c		650a		5765		2077		696c		6c20		6c6f		6f6b	
0000040	sp	a	t	sp	t	h	i	s	sp	f	i	l	e	sp	u	n
	2061		7420		7468		6973		2066		696c		6520		756e	
0000060	d	e	r	sp	U	n	i	x	sp	a	n	d	sp	D	O	S
	6465		7220		556e		6978		2061		6e64		2044		4f53	
0000100	nl															
	0a00															
0000101																

FIGURE 1.2 Memory Implementation for Variable t

| 41 |
| 42 |
| 43 |
| 44 |
| 45 |
| 46 |
| 47 |
| 48 |

Note: This is a particular implementation for a given machine. A different machine might elect to store the data differently. The important part is that the differences be transparent to the user.

All values are in hex.

FIGURE 1.3 Mapping of each Union Entry

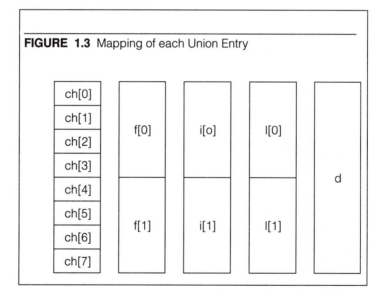

The organization of each union entry is shown in Figure 1.3. For the union declaration *t* there are only eight bytes stored in memory. These eight bytes can be interpreted as eight individual characters or two longs or two doubles,

Code List 1.15 Data Representations

C++ Source
| ```
#include <iostream.h>
union test
{
double d;
float f[2];
char ch[8];
int i[2];
long l[2];
};
void main()
{
union test t;
int i;
cout << "The size of double is " << sizeof(double) << endl;
cout << "The size of float is " << sizeof(float) << endl;
cout << "The size of char is " << sizeof(char) << endl;
cout << "The size of int is " << sizeof(int) << endl;
cout << "The size of long is " << sizeof(long) << endl;
t.l[0]=0x41424344;
t.l[1]=0x45464748;
cout << "As characters: "; for(i=0;i<8;i++) cout << t.ch[i]; cout << endl;
cout << "As a double: " << t.d << endl;
cout << "As two integers: " << t.i[0] << " " << t.i[1] << endl;
cout << "As two longs: " << t.l[0] << " " << t.l[1] << endl;
cout << "As two floats: " << t.f[0] << " " << t.f[1] << endl;
}
``` |

Code List 1.16 Output of Program in Code List 1.15

C++ Output
The size of double is 8 The size of float is 4
```

Code List 1.16 Output of Program in Code List 1.15 (continued)

C++ Output
The size of char is 1
The size of int is 4
The size of long is 4
As characters: ABCDEFGH
As a double: 2.39374e+06
As two integers: 1094861636 1162233672
As two longs: 1094861636 1162233672
As two floats: 12.1414 3172.46

etc. For instance by looking at Table 1.8 one sees the value of *ch[0]* which is 0x41 which is the letter A. Similarly, the value of *ch[1]* is 0x42 which is the letter B. When interpreted as an integer the value of *i[0]* is 0x41424344 which is in 2's complement format. Converting to decimal one has *i[0]* with the value of

$$i\,[0] \;=\; 68 + 67\,(256) + 66\,(256^2) + 65\,(256^3) \;=\; 1094861636 \qquad \textbf{(1.49)}$$

If one were to interpret 0x41424344 as an IEEE 32-bit floating point number its value would be 12.1414. If one were to interpret 0x45464748 as an IEEE 32-bit floating point number its value would be 3172.46.

There are only one's and zero's stored in memory and collections of bits can be interpreted to be characters or integers or floating point numbers. To determine which kind of operations to perform the compiler must be able to determine the type of each operation.

1.5 Problems

(1.1) Represent the following decimal numbers when possible in the format specified. 125, –1000, 267, 45, 0, 2500. Generate all answers in HEX!

  . a) 8-bit 2's complement—2 hex digits
     b) 16-bit 2's complement—4 hex digits
     c) 32-bit 2's complement—8 hex digits
     d) 64-bit 2's complement—16 hex digits

(1.2) Convert the 12-bit 2's complement numbers that follows to 32-bit 2's complement numbers. Present your answer with 8 hex digits.

a) 0xFA4

b) 0x802

c) 0x400

d) 0x0FF

(1.3) Represent decimal 0.35 in IEEE 32-bit format and IEEE 64-bit format.

(1.4) Represent the decimal fraction $4/7$ in binary.

(1.5) Represent the decimal fraction 0.3 in octal.

(1.6) Represent the decimal fraction 0.85 in hex.

(1.7) Calculate the floating point number represented by the IEEE 32-bit representation F8080000.

(1.8) Calculate the floating point number represented by the IEEE 64-bit representation F808000000000000.

(1.9) Write down the ASCII representation for the string "Hello, how are you?". Strings in C++ are terminated with a 00 in hex (a null character). Terminate your string with the null character. Do not represent the quotes in your string. The quotes in C++ are used to indicate the enclosure is a string.

(1.10) Write a C++ program that outputs "Hello World".

(1.11) In Code List 1.8 the twos complement of the largest representable negative integer, -32768, is the same number. Explain this result. Is the theory developed incorrect?

(1.12) In Section 1.1.4 the issue of conversion is assessed for signed-magnitude, unsigned, and 2's complement numbers. Is there a simple algorithm to convert an IEEE 32-bit floating point number to IEEE 64-bit floating point number?

2 Algorithms

This chapter presents the fundamental concepts for the analysis of algorithms.

2.1 Order

N denotes the set of natural numbers, $\{1, 2, 3, 4, 5, \ldots\}$.

Definition 2.1

A sequence, x, over the real numbers is a function from the natural numbers into the real numbers:

$$x : N \rightarrow R$$

x_1 is used to denote the first element of the sequence, $x(1)$. In general,

$$x = \{x(1), x(2), \ldots, x(n), \ldots\}$$

and will be written as

$$x = x_1, x_2, \ldots, x_n, \ldots \tag{2.1}$$

❐

Unless otherwise noted, when x is a sequence and f is a function of one variable, $f(x)$, is the sequence obtained by applying the function f to each of the elements of x. If

$$y = f(x)$$

then

$$y_k = f(x_k)$$

For example,

$$|x| = |x_1|, |x_2|, ..., |x_n|, ...$$

$$3x = 3x_1, 3x_2, ..., 3x_n, ...$$

Definition 2.2

If x and y are sequences, then x is of order at most y, written $x \in O(y)$, if there exists a positive integer N and a positive number k such that

$$x_n \leq ky_m \qquad \text{for all } n > N \qquad \text{(2.2)}$$

❐

Definition 2.3

If x and y are sequences then x is of order exactly y, written, $x \in \Theta(y)$, if $x \in O(y)$ and $y \in O(x)$.

❐

Definition 2.4

If x and y are sequences then x is of order at least y, written, $x \in \Omega(y)$, if $y \in O(x)$.

❐

Definition 2.5

The time complexity of an algorithm is the sequence

$$t = t_1, t_2, ...$$

where t_k is the number of time steps required for solution of a problem of size k.

❐

EXAMPLE 2.1 Time Complexity

Find the time complexity sequence for the addition of $2k$ numbers. Assume an infinite number of processors are available and each processor is capable of performing the addition of two numbers in a single time step.

Solution: If addition is performed in a tree-like manner, the time for computation with k processors is

$$t_k = \lceil \log_2 2k \rceil = 1 + \lceil \log_2 k \rceil$$

$$t = \{1, 2, 3, 3, 4, 4, 4, 4, 5, 5, 5, 5, 5, 5, 5, 5, 6, 6, \dots\}$$

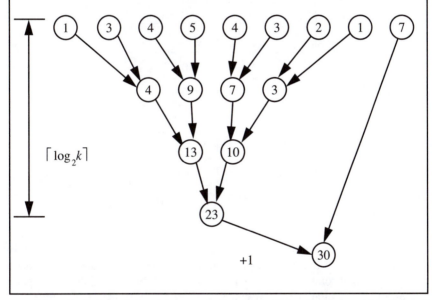

The calculation of the time complexity for addition is illustrated in Example 2.1. A comparison of the order of several classical functions is shown in Table 2.1. The time required for a variety of operations on a 100 Megaflop machine

TABLE 2.1 Order Comparison

Function	n=1	n=10	n=100	n=1000	n=10000
$\log(n)$	0	3.32	6.64	9.97	13.3
$n\log(n)$	0	33.2	664	9.97×10^3	1.33×10^5

TABLE 2.1   Order Comparison (continued)

Function	n=1	n=10	n=100	n=1000	n=10000
n^2	1	100	10000	1×10^6	1×10^8
n^5	1	1×10^5	1×10^{10}	1×10^{15}	1×10^{20}
e^n	2.72	2.2×10^4	2.69×10^{43}	1.97×10^{434}	8.81×10^{4342}
$n!$	1	3.63×10^6	9.33×10^{157}	4.02×10^{2567}	2.85×10^{35659}

is illustrated in Table 2.2. As can be seen from Table 2.1 if a problem is truly

TABLE 2.2   Calculations for a 100 MFLOP machine

Time	# of Operations
1 second	10^8
1 minute	$6x10^9$
1 hour	$3.6x10^{11}$
1 day	$8.64x10^{12}$
1 year	$3.1536x10^{15}$
1 century	$3.1536x10^{17}$
100 trillion years	$3.1536x10^{29}$

of exponential order then it is unlikely that a solution will ever be rendered for
the case of n=100. It is this fact that has led to the use of heuristics in order to
find a "good solution" or in some cases "a solution" for problems thought to
be of exponential order. An example of Order is shown in Example 2.2.
through Example 2.4.

2.1.1   Justification of Using Order as a Complexity Measure

One of the major motivations for using Order as a complexity measure is to
get a handle on the inductive growth of an algorithm. One must be extremely
careful however to understand that the definition of Order is "in the limit." For
example, consider the time complexity functions f_1 and f_2 defined in Example
2.6. For these functions the asymptotic behavior is exhibited when $n \geq 10^{50}$.
Although $f_1 \in \Theta(e^n)$ it has a value of 1 for $n < 10^{50}$. In a pragmatic sense it

---

EXAMPLE 2.2 Order

Show that $n\log n \in \Theta\left(\log\left(n!\right)\right)$

Solution:

$$\log\left(n!\right) = \log\left(1 \times 2 \times \ldots n\right)$$
$$= \log\left(1\right) + \log\left(2\right) + \ldots \log\left(n\right)$$
$$\leq \log\left(n\right) + \log\left(n\right) + \ldots \log\left(n\right)$$
$$= n\log n$$

so

$$n\log n \in \Omega\left(\log\left(n!\right)\right)$$

Similarly

$$\log\left(n!\right) \geq \log\left(\frac{n}{2}\right) + \log\left(\frac{n}{2} + 1\right) + \ldots + \log\left(n\right)$$

$$\log\left(n!\right) \geq \frac{n}{2}\log\left(\frac{n}{2}\right)$$

$$\log\left(n!\right) \geq \frac{n}{2}\log\left(n\right) - \frac{n}{2}\log\left(2\right)$$

$$\log\left(n!\right) \geq \frac{n\log\left(n\right)}{4} \qquad n > 10$$

so

$$n\log\left(n\right) \in O\left(\log\left(n!\right)\right)$$

---

would be desirable to have a problem with time complexity f_1 rather than f_2. Typically, however, this phenomenon will not appear and generally one might assume that it is better to have an algorithm which is $\Theta\left(1\right)$ rather than

---

EXAMPLE 2.3 Order

Find a sequence f such that

$$f \notin O(n) \quad \text{and} \quad f \notin \Omega(n)$$

Solution:

One possible instance is

$$f(n) = \begin{cases} \sqrt{n}, \text{n odd} \\ n^2, \text{n even} \end{cases}$$

$\Theta(e^n)$. One should always remember that the constants of order can be significant in real problems.

2.2 Induction

Simple induction is a two step process:

- Establish the result for the case $N = 1$
- Show that if is true for the case $N = n$ then it is true for the case $N = n + 1$

This will establish the result for all $n > 1$.

Induction can be established for any set which is well ordered. A well-ordered set, S, has the property that if

$$x, y \in S$$

then either

- $x < y$
- $x > y$ or
- $x = y$

EXAMPLE 2.4 Order

If $f \in O(g)$ and $g \in O(h)$ prove that $f \in O(h)$

Solution: From the definition of order there exists N_1, N_2, k_1, k_2 such that

$$f_n \leq k_1 g_n \qquad n > N_1$$

and

$$g_n \leq k_2 h_n \qquad n > N_2$$

therefore,

$$f_n \leq k_1 k_2 h_n \qquad n > max(N_1, N_2)$$

with $k_3 = k_1 k_2$ and $N_3 = max(N_1, N_2)$

one has

$$f_n \leq k_3 h_n \qquad n > N_3$$

which gives the desired result. ❐

Additionally, if S' is a nonempty subset of S:

$$S' \subseteq S \qquad S' \neq \emptyset$$

then S' has a least element. An example of simple induction is shown in Example 2.5.

The well-ordering property is required for the inductive property to work. For example consider the method of *infinite descent* which uses an inductive type approach. In this method it is required to demonstrate that a specific property cannot hold for a positive integer. The approach is as follows:

EXAMPLE 2.5 Induction

Show using induction that

$$\sum_{j=0}^{n} \frac{1}{(1+j)(2+j)} = 1 - \frac{1}{n+2}$$

Step 1: Establish the case for $n = 0$.

$$\frac{1}{(1+0)(2+0)} = 1 - \frac{1}{0+2}$$

Step 2: Assume true for n and establish the case for $n+1$.

Let

$$f(n) = \sum_{j=0}^{n} \frac{1}{(1+j)(2+j)}$$

then

$$f(n+1) = f(n) + \frac{1}{(1+n+1)(2+n+1)}$$

$$= 1 - \frac{1}{n+2} + \frac{1}{(n+2)(n+3)}$$

$$= 1 + \frac{1 - (n+3)}{(n+2)(n+3)}$$

$$= 1 - \frac{1}{n+3}$$

$$= 1 - \frac{1}{(n+1)+2}$$

1. Let $P(k) = TRUE$ denote that a property holds for the value of k. Also assume that $P(0)$ does not hold so $P(0) = FALSE$.

Let S be the set that

$$S = \{k : P(k) = TRUE\} \qquad k = 1, 2, 3, \ldots \qquad (2.3)$$

From the well-ordering principle it is true that if S is not empty then S has a smallest member. Let j be such a member:

$$j = \min_{k} (P(k) = TRUE) \qquad (2.4)$$

2. Prove that $P(j)$ implies $P(j-1)$ and this will lead to a contradiction since $P(0)$ is $FALSE$ and j was assumed to be minimal so that S must be empty. This implies the property does not hold for any positive integer k. See Problem 2.1 for a demonstration of *infinite descent*.

2.3 Recursion

Recursion is a powerful technique for defining an algorithm.

Definition 2.6

A procedure is *recursive* if it is, whether directly or indirectly, defined in terms of itself.

❏

2.3.1 Factorial

One of the simplest examples of recursion is the factorial function $f(n) = n!$. This function can be defined recursively as

$$f(0) = 1 \qquad (2.5)$$

$$f(n) = nf(n-1) \qquad n > 0 \qquad (2.6)$$

A simple C++ program implementing the factorial function recursively is shown in Code List 2.1. The output of the program is shown in Code List 2.2.

Code List 2.1 Factorial

```
C++ Source Program

#include <iostream.h>
double fact(double x)
{
if(x==1.0) return(1.0);
else return(x*fact(x–1.0));
}


main()
{
int i;
for(i=1;i<10;i++) cout << fact(i) << endl;
}
```

Code List 2.2 Output of Program in Code List 2.1

```
C++ Output

1
2
6
24
120
720
5040
40320
362880
```

2.3.2 Fibonacci Numbers

The Fibonacci sequence, $F(n)$, is defined recursively by the recurrence relation

$$F(n) = F(n-1) + F(n-2) \qquad \text{(2.7)}$$

$$F(0) = 0 \qquad F(1) = 1 \qquad \text{(2.8)}$$

A simple program which implements the Fibonacci sequence recursively is shown in Code List 2.3. The output of the program is shown in   Code List 2.4.

Code List 2.3 Fibonacci Sequence Generation

```
C++ Source Code

#include <iostream.h>
#include <math.h>


int f(int x)
{
if(x>1) return f(x–1)+f(x–2);
else if(x==1) return 1;
else return 0;
}


void main()
{
int x;
for(x=0;x<20;x++)
        {
        cout << "The value for " << x << " is " << f(x) << endl;
        }
}
```

Code List 2.4 Output of Program in Code List 2.3

```
C++ Output

The value for 0 is 0
The value for 1 is 1
The value for 2 is 1
The value for 3 is 2
The value for 4 is 3
The value for 5 is 5
The value for 6 is 8
```

Code List 2.4 Output of Program in Code List 2.3 (continued)

C++ Output
The value for 7 is 13
The value for 8 is 21
The value for 9 is 34
The value for 10 is 55
The value for 11 is 89
The value for 12 is 144
The value for 13 is 233
The value for 14 is 377
The value for 15 is 610
The value for 16 is 987
The value for 17 is 1597
The value for 18 is 2584
The value for 19 is 4181

The recursive implementation need not be the only solution. For instance in looking for a closed solution to the relation if one assumes the form $F(n) = \lambda^n$ one has

$$\lambda^n = \lambda^{n-1} + \lambda^{n-2} \tag{2.9}$$

which assuming $\lambda \neq 0$

$$\lambda^2 = \lambda + 1 \tag{2.10}$$

The solution via the quadratic formula yields

$$\lambda = \frac{1 \pm \sqrt{5}}{2} \tag{2.11}$$

Because Eq. 2.7 is linear it admits solutions of the form

$$F(n) = A\left(\frac{1 + \sqrt{5}}{2}\right)^n + B\left(\frac{1 - \sqrt{5}}{2}\right)^n \tag{2.12}$$

To satisfy the boundary conditions in Eq. 2.8 one obtains the matrix form

$$\begin{bmatrix} 1 & 1 \\ \dfrac{1+\sqrt{5}}{2} & \dfrac{1-\sqrt{5}}{2} \end{bmatrix} \begin{bmatrix} A \\ B \end{bmatrix} = \begin{bmatrix} 0 \\ 1 \end{bmatrix} \qquad (2.13)$$

multiplying both sides by the 2×2 matrix inverse

$$\begin{bmatrix} A \\ B \end{bmatrix} = \dfrac{-1}{\sqrt{5}} \begin{bmatrix} \dfrac{1-\sqrt{5}}{2} & -1 \\ -\dfrac{1+\sqrt{5}}{2} & 1 \end{bmatrix} \begin{bmatrix} 0 \\ 1 \end{bmatrix} \qquad (2.14)$$

which yields

$$A = \dfrac{\sqrt{5}}{5} \qquad (2.15)$$

$$B = -\dfrac{\sqrt{5}}{5} \qquad (2.16)$$

resulting in the closed form solution

$$F(n) = \dfrac{\sqrt{5}}{5} \left(\left(\dfrac{1+\sqrt{5}}{2} \right)^n - \left(\dfrac{1-\sqrt{5}}{2} \right)^n \right) \qquad (2.17)$$

A nonrecursive implementation of the Fibonacci series is shown in Code List 2.5. The output of the program is the same as the recursive program given in Code List 2.4.

Code List 2.5 Fibonacci Program — Non Recursive Solution

C++ Source Code
```#include <iostream.h>```
```#include <math.h>```
```void main()```
```{```
```double x,y;```
```for(y=0.0;y<20.0;y++)```
```   {```

Code List 2.5 Fibonacci Program — Non Recursive Solution (continued)

C++ Source Code

x=sqrt(5.0)/5.0*pow(0.5+sqrt(5.0)/2.0,y);
 x-=sqrt(5.0)/5.0*pow(0.5-sqrt(5.0)/2.0,y);
 cout << "The value for " << y << " is " << (int) (x+0.5) << endl;
 }

}

2.3.3 General Recurrence Relations

This section presents the methodology to handle general 2nd order recurrence relations. The recurrence relation given by

$$aR(n) = bR(n-1) + cR(n-2) \tag{2.18}$$

with initial conditions:

$$R(0) = d \qquad R(1) = e \tag{2.19}$$

can be solved by assuming a solution of the form $R(n) = \lambda^n$. This yields

$$a\lambda^2 - b\lambda - c = 0 \tag{2.20}$$

If the equation has two distinct roots, λ_1, λ_2, then the solution is of the form

$$R(n) = C_1\lambda_1^n + C_2\lambda_2^n \tag{2.21}$$

where the constants, C_1, C_2, are chosen to enforce Eq. 2.19. If the roots, however, are not distinct then an alternate solution is sought:

$$R(n) = C_1 n\lambda^n + C_2\lambda^n \tag{2.22}$$

where λ is the double root of the equation. To see that the term $C_1 n\lambda^n$ satisfies the recurrence relation one should note that for the multiple root Eq. 2.18 can be written in the form

$$R(n) = 2\lambda R(n-1) - \lambda^2 R(n-2) \tag{2.23}$$

Substituting $C_1 n\lambda^n$ into Eq. 2.23 and simplifying verifies the solution.

2.3.4 Tower of Hanoi

The Tower of Hanoi problem is illustrated in Figure 2.1. The problem is to move n discs (in this case, three) from the first peg, A, to the third peg, C. The middle peg, B, may be used to store discs during the transfer. The discs have to be moved under the following condition: at no time may a disc on a peg have a wider disc above it on the same peg. As long as the condition is met all three pegs may be used to complete the transfer. For example the problem may be solved for the case of three by the following move sequence:

$$(A, C), (A, B), (C, B), (A, C), (B, A), (B, C), (A, C) \qquad \text{(2.24)}$$

where the ordered pair, (x, y), indicates to take a disk from peg x and place it on peg y.

FIGURE 2.1 Tower of Hanoi Problem

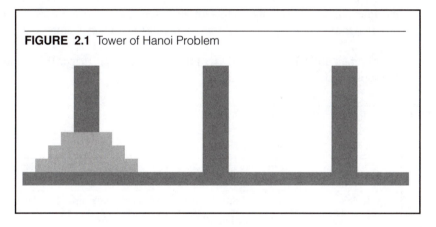

The problem admits a nice recursive solution. The problem is solved in terms of n by noting that to move n discs from A to C one can move $n-1$ discs from A to B move the remaining disc from A to C and then move the $n-1$ discs from B to C. This results in the relation for the number of steps, $S(n)$, required for size n as

$$S(n) = 2S(n-1) + 1 \qquad \text{(2.25)}$$

with the boundary conditions

$$S(1) = 1 \qquad S(2) = 3 \qquad \text{(2.26)}$$

Eq. 2.25 admits a solution of the form

$$S(n) = A2^n + B \qquad (2.27)$$

and matching the boundary conditions in Eq. 2.26 one obtains

$$S(n) = 2^n - 1 \qquad (2.28)$$

A growing field of interest is the visualization of algorithms. For instance, one might want to animate the solution to the Tower of Hanoi problem. Each disc move results in a new picture in the animation. If one is to incorporate the pictures into a document then a suitable language for its representation is Post-Script[1]. This format is supported by almost all word processors and as a result is encountered frequently. A program to create the PostScript® description of the Tower of Hanoi is shown in Code List 2.6 The program creates an encapsulated postscript file shown in Code List 2.7. The word processor used to generate this book took the output of the program in Code List 2.7 and imported it to yield Figure 2.1! This program illustrates many features of C++.

The program utilizes only a small set of the PostScript® language. This primitive subset is described in Table 2.3.

TABLE 2.3 PostScript® — Primitive Subset

Command	Description
x setgray	set the gray level to x. $x = 1$ is white and $x = 0$ is black. This will affect the fill operation.
x y scale	scale the X dimension by x and scale the Y dimension by y.
x setlinewidth	set the linewidth to x.
x y moveto	start a subpath and move to location x y on the page.
x y rlineto	draw a line from current location (x_1, y_1) to $(x_1 + x, y_1 + y)$. Make the endpoint the current location. Appends the line to the subpath.
fill	close the subpath and fill the area enclosed.
newpath	create a new path with no current point.
showpage	displays the page to the output device.

The program uses a number of classes in C++ which are *derived* from one another. This is one of the most powerful concepts in object-oriented programming. The class structure is illustrated in Figure 2.2.

1. PostScript® is a trademark of Adobe Systems Inc.

In the figure there exists a high-level base class called the graphic context. In a typical application a number of subclasses might be derived from it. In this case the graphics context specifies the line width, gray scale, and scale for its subsidiary objects. A derived class from the graphics context is the object class. This class contains information about the position of the object. This attribute is common to objects whether they are rectangles, circles, etc. A derived class from the object class is the rectangle class. For this class, specific information about the object is kept which identifies it with a rectangle, namely the width and the height. The draw routine overrides the virtual draw function for the object. The *draw* function in the object class is *void* even though for more complex examples it might have a number of operations. The RECTANGLE class inherits all the functions from the GRAPHICS_CONTEXT class and the OBJECT class.

In the program, the rectangle class instantiates the discs, the base, and the pegs. Notice in Figure 2.1 that the base and pegs are drawn in a different gray scale than the discs. This is accomplished by the two calls in *main()*:

- *peg.set_gray(0.6)*
- *base.set_gray(0.6)*

Any object of type RECTANGLE defaults to a *set_gray* of 0.8 as defined in the constructor function for the rectangle. Notice that peg is declared as a RECTANGLE and has access to the *set_gray* function of the GRAPHICS_- CONTEXT. The valid operations on *peg* are:

- *peg.set_line_width()*, from the GRAPHICS_CONTEXT class
- *peg.set_scale()*, from the GRAPHICS_CONTEXT class
- *peg.set_gray()*, from the GRAPHICS_CONTEXT class
- *peg.location()*, from the OBJECT class
- *peg.set_location()*, from the RECTANGLE class
- *peg.set_width()*, from the RECTANGLE class
- *peg.set_height()*, from the RECTANGLE class
- *peg.draw()*, from the RECTANGLE class

The virtual function *draw* in the OBJECT class is hidden from *peg* but it can be accessed in C++ using the scoping operator with the following call:

- *peg.object::draw()*, uses *draw* from the OBJECT class

Hence, in the program, all the functions are available to each instance of the rectangle created. This availability arises because the functions are declared as public in each class and each derived class is also declared public. Without the public declarations C++ will hide the functions of the base class from the derived class. Similarly, the data the functions access are declared as protected which makes the data visible to the functions of the derived classes.

The first peg in the program is created with rectangle *peg(80,0,40,180)*. The gray scale for this peg is changed from the default of 0.8 to 0.6 with *peg.set_-gray(0.6)*. The peg is drawn to the file with *peg.draw(file)*. This draw operation results in the following lines placed in the file:

- newpath
- 1 setlinewidth
- 0.6 setgray
- 80 0 moveto
- 0 180 rlineto
- 40 0 rlineto
- 0 −180 rlineto
- fill

The PostScript® action taken by the operation is summarized in Figure 2.3. Note that the rectangle in the figure is not drawn to scale. The drawing of the base and the discs follows in an analogous fashion.

Code List 2.6 Program to Display Tower of Hanoi

C++ Source
| ```
#include <iostream.h>
#include <iomanip.h>
#include <fstream.h>
//This program creates an encapsulated postscript file
//to draw the Tower of Hanoi
class graphics_context {
        protected:
                double line_w;
                double x_scale, y_scale;
                double gray;
        public:
                void set_line_width(double x=1.0) {line_w=x;};
                void set_scale(double x=1.0, double y=1.0)
                        {x_scale=x;y_scale=y;};
                void set_gray(double x=0.0) {gray=x;};
``` |
```

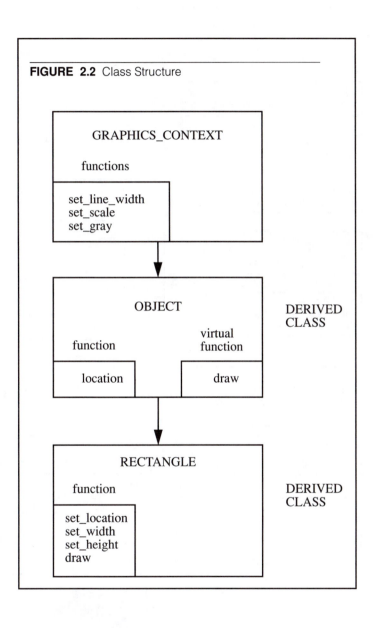

FIGURE 2.2 Class Structure

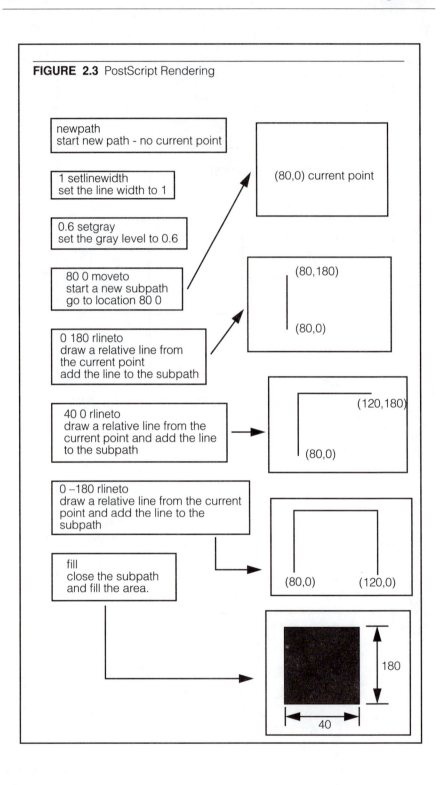

FIGURE 2.3 PostScript Rendering

Code List 2.6 Program to Display Tower of Hanoi (continued)

C++ Source

```cpp
};
class object : public graphics_context {
        protected:
        double x_loc, y_loc;
        public:
        void location(double x=0.0, double y=0.0) { x_loc=x; y_loc=y; };
        virtual void draw() { };
};
class rectangle : public object {
        protected:
        double width, height;
        public:
        rectangle(double x=0.0, double y=0.0, double w=1.0,
                double h=1.0) { x_loc=x; y_loc=y; width=w; height=h;
                        set_scale();
                        set_line_width();
                        set_gray(0.8);};
        void set_location(double x, double y) { x_loc=x; y_loc=y; };
        void set_width(double w) { width=w; };
        void set_height(double h) { height=h; };
        void draw(ofstream& file)
                {
                file << "newpath" << endl;
                file << line_w << " setlinewidth" << endl;
                file << gray << " setgray" << endl;
                file << x_loc << " " << y_loc << " moveto" << endl;
                file << "0 " << height*y_scale << " rlineto" << endl;
                file << width*x_scale << " 0 rlineto" << endl;
                file << "0 " << -height*y_scale << " rlineto" << endl;
                file << "fill" << endl;
                };
        };
```

Code List 2.6 Program to Display Tower of Hanoi (continued)

C++ Source

```
void main()
{
        ofstream file("tower.eps",ios::outlios::trunc);
        if(!file)
                {
                cout << "Could not open file\n";
                return;
                }
        // Add standard postscript header
        file << "%!PS-Adobe-2.0 EPSF-2.0" << endl;
        file << "%%BoundingBox: 0 0 300 90" << endl;
        file << "%%Creator: Alan Parker" << endl;
        file << "%%EndComments" << endl;
        file << "0.8 setgray" << endl;
        file << "0.5 0.5 scale" << endl;
        // create the first peg and draw it
        rectangle peg(80,0,40,180);
        peg.set_gray(0.6);
        peg.draw(file);
        peg.set_location(280,0);
        peg.draw(file);
        peg.set_location(480,0);
        peg.draw(file);
        // create the base
        rectangle base(0,0,600,20);
        base.set_gray(0.6);
        base.draw(file);
        // create the disc and draw it
        rectangle disc(20,20,160,20);
        disc.draw(file);
        disc.set_location(40,40);
        disc.set_width(120);
```

Code List 2.6 Program to Display Tower of Hanoi (continued)

C++ Source
| ```
        disc.draw(file);
        disc.set_location(60,60);
        disc.set_width(80);
        disc.draw(file);
        // Close file with standard trailer
        file << "showpage" << endl << "%%Trailer" << endl;
        file.close();
}
``` |

Code List 2.7 File Created by Program in Code List 2.6

File Tower.eps
%!PS-Adobe-2.0 EPSF-2.0
%%BoundingBox: 0 0 300 90
%%Creator: Alan Parker
%%EndComments
0.8 setgray
0.5 0.5 scale
newpath
1 setlinewidth
0.6 setgray
80 0 moveto
0 180 rlineto
40 0 rlineto
0 −180 rlineto
fill
newpath
1 setlinewidth
0.6 setgray
280 0 moveto
0 180 rlineto
40 0 rlineto
```

Code List 2.7 File Created by Program in Code List 2.6 (continued)

File Tower.eps
0 −180 rlineto
fill
newpath
1 setlinewidth
0.6 setgray
480 0 moveto
0 180 rlineto
40 0 rlineto
0 −180 rlineto
fill
newpath
1 setlinewidth
0.6 setgray
0 0 moveto
0 20 rlineto
600 0 rlineto
0 −20 rlineto
fill
newpath
1 setlinewidth
0.8 setgray
20 20 moveto
0 20 rlineto
160 0 rlineto
0 −20 rlineto
fill
newpath
1 setlinewidth
0.8 setgray
40 40 moveto
0 20 rlineto
120 0 rlineto

Code List 2.7 File Created by Program in Code List 2.6(continued)

File Tower.eps
0 –20 rlineto
fill
newpath
1 setlinewidth
0.8 setgray
60 60 moveto
0 20 rlineto
80 0 rlineto
0 –20 rlineto
fill
showpage
%%Trailer

2.3.5 Boolean Function Implementation

This section presents a recursive solution to providing an upper bound to the number of 2-input NAND gates required to implement a boolean function of n boolean variables. The recursion is obtained by noticing that a function, $f(x_1, x_2, ..., x_n)$ of n variables can be written as

$$f(x_1, x_2, ..., x_n) = x_n g(x_1, ..., x_{n-1}) + \overline{x_n} h(x_1, ..., x_{n-1}) \qquad (2.29)$$

for some functions g and h of $n-1$ boolean variables. The implementation is illustrated in Figure 2.4.

The number of NAND gates thus required as a function of n, $C(n)$, can be written recursively as:

$$C(n) = 2C(n-1) + 4 \qquad (2.30)$$

The solution to the simple recurrence relation yields, assuming a general form of $C(n) = \lambda^n$ followed by a constant to obtain the particular solution

$$C(n) = A2^n + B \qquad (2.31)$$

Applying the boundary condition $C(1) = 1$ and $C(2) = 6$ one obtains

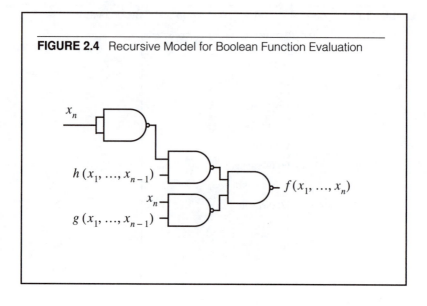

FIGURE 2.4 Recursive Model for Boolean Function Evaluation

$$C(n) = 5(2^n) - 4 \qquad (2.32)$$

2.4 Graphs and Trees

This section presents some fundamental definitions and properties of graphs.

Definition 2.7

A *graph* is a collection of vertices, V, and associated edges, E, given by the pair

$$G = (V, E) \qquad (2.33)$$

❏

A simple graph is shown in Figure 2.5.

In the figure the graph shown has

$$V = \{v_1, v_2, v_3\} \qquad (2.34)$$

$$E = \{(v_1, v_2), (v_2, v_3), (v_3, v_1)\} \qquad (2.35)$$

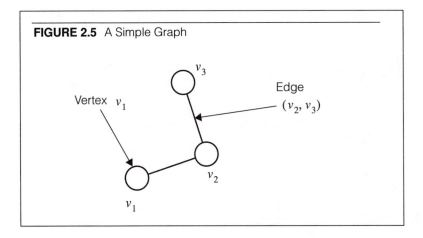

FIGURE 2.5 A Simple Graph

Definition 2.8

The *size* of a graph is the number of edges in the graph

$$size\,(G) \;=\; |E| \tag{2.36}$$

❐

Definition 2.9

The *order* of a graph G is the number of vertices in a graph

$$order\,(G) \;=\; |V| \tag{2.37}$$

❐

For the graph in Figure 2.5 one has

$$size\,(G) \;=\; 2 \qquad order\,(G) \;=\; 3 \tag{2.38}$$

Definition 2.10

The *degree* of a vertex (also referred to as a node), in a graph, is the number of edges containing the vertex.

❐

Definition 2.11

In a graph, $G = (V, E)$, two vertices, v_1 and v_2, are *neighbors* if

$$(v_1, v_2) \in E \text{ or } (v_2, v_1) \in E$$

❏

In the graph in Figure 2.5 v_1 and v_2 are neighbors but v_1 and v_3 are not neighbors.

Definition 2.12

If $G = (V_1, E_1)$ is a graph, then $H = (V_2, E_2)$ is a *subgraph* of G written $H \subseteq G$ if $V_2 \subseteq V_1$ and $E_2 \subseteq E_1$.

❏

A subgraph of the graph in Figure 2.5 is shown in Figure 2.6.

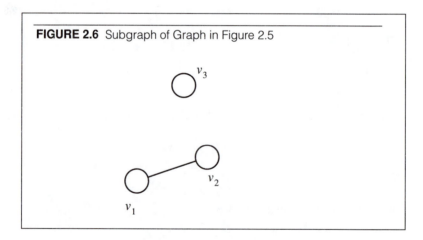

FIGURE 2.6 Subgraph of Graph in Figure 2.5

The subgraph is generated from the original graph by the deletion of a single edge (v_2, v_3).

Definition 2.13

A *path* is a collection of neighboring vertices.

❐

For the graph in Figure 2.5 a valid path is

$$path = (v_1, v_2, v_3) \tag{2.39}$$

Definition 2.14

A graph is *connected* if for each vertex pair (v_i, v_j) there is a path from v_i to v_j.

❐

The graph in Figure 2.5 is connected while the graph in Figure 2.6 is disconnected.

Definition 2.15

A *directed graph* is a graph with vertices and edges where each edge has a specific direction relative to each of the vertices.

❐

An example of a directed graph is shown in Figure 2.7.

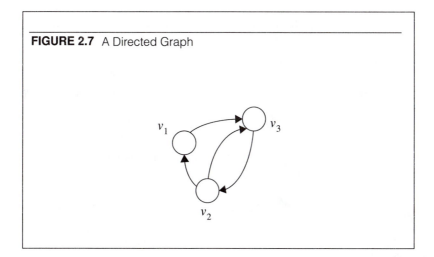

FIGURE 2.7 A Directed Graph

The graph in the figure has $G = (V, E)$ with

$$V = \{v_1, v_2, v_3\} \tag{2.40}$$

$$E = \{(v_1, v_2), (v_2, v_3), (v_3, v_2), (v_2, v_1)\} \tag{2.41}$$

In a directed graph the edge (v_i, v_j) is not the same as the edge (v_j, v_i) when $i \neq j$. The same terminology $G = (V, E)$ will be used for directed and undirected graphs; however, it will always be stated whether the graph is to be interpreted as a directed or undirected graph.

The definition of path applies to a directed graph also. As shown in Figure 2.8 there is a path from v_1 to v_4 but there is no path from v_2 to v_5.

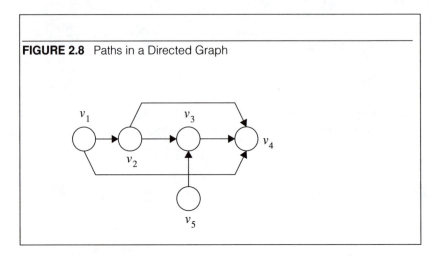

FIGURE 2.8 Paths in a Directed Graph

A number of paths exist from v_1 to v_4, namely

$$p_1 = (v_1, v_2, v_3, v_4) \qquad p_2 = (v_1, v_2, v_4) \qquad p_3 = (v_1, v_4) \tag{2.42}$$

Definition 2.16

A *cycle* is a path from a vertex to itself which does not repeat any vertices except the first and the last.

❐

A graph containing no cycles is said to be acyclic. An example of cyclic and acyclic graphs is shown in Figure 2.9.

FIGURE 2.9 Cyclic and Acyclic Graphs

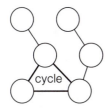

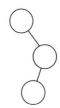

An Undirected Cyclic Graph   An Undirected Acyclic Graph

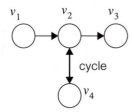

A Directed Acyclic Graph
(DAG)

A Directed Cyclic Graph

Notice for the directed cyclic graph in Figure 2.9 that the double arrow notations between nodes v_2 and v_4 indicate the presence of two edges (v_2, v_4) and (v_4, v_2). In this case it is these edges which form the cycle.

Definition 2.17

A *tree* is an acyclic connected graph.

◻

Examples of trees are shown in Figure 2.10.

Definition 2.18

An edge, e, in a connected graph, $G = (V, E)$, is a *bridge* if $G' = (V, E')$ is disconnected where

$$E' = E - e \tag{2.43}$$

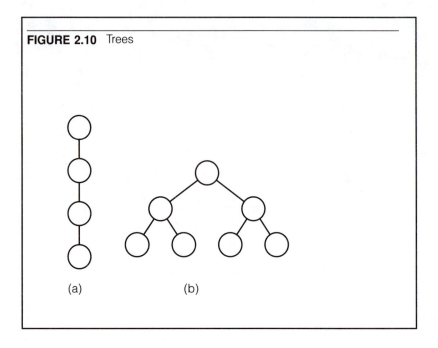

FIGURE 2.10 Trees

(a)   (b)

❒

If the edge, e, is removed, the graph, G, is divided into two separate connected graphs. Notice that every edge in a tree is a bridge.

Definition 2.19

A *planar* graph is a graph that can be drawn in the plane without any edges intersecting.

❒

An example of a planar graph is shown in Figure 2.11. Notice that it is possible to draw the graph in the plane with edges that cross although it is still planar.

Definition 2.20

The *transitive closure* of a directed graph, $G = (V_1, E_1)$ is a graph, $H = (V_2, E_2)$, such that,

$$V_2 = V_1 \qquad\qquad\qquad (2.44)$$

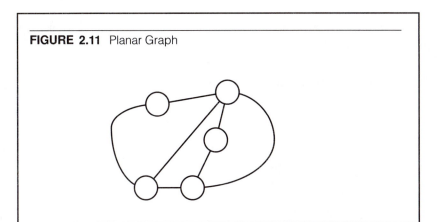

$$E_2 = f(V_1, E_1) \qquad \text{(2.45)}$$

where f returns a set of edges. The set of edges is as follows:

$$(v_1, v_2) \in f(V_1, E_1) \qquad \text{if there is a path from } v_1 \text{ to } v_2 \qquad \text{(2.46)}$$

❐

Thus in Eq. 2.45, $E_2 \supseteq E_1$. Transitive closure is illustrated in Figure 2.12.

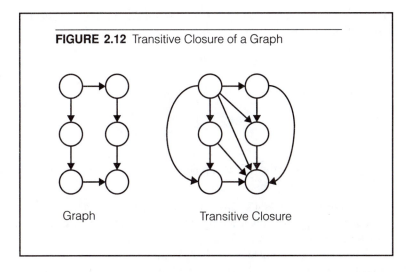

2.5 Parallel Algorithms

This section presents some fundamental properties and definitions used in parallel processing.

2.5.1 Speedup and Amdahls Law

Definition 2.21

The speedup of an algorithm executed using n parallel processors is the ratio of the time for execution on a sequential machine, T_{SEQ}, to the time on the parallel machine, T_{PAR}:

$$\text{Speedup}\,(n) \;=\; \frac{T_{SEQ}}{T_{PAR}} \tag{2.47}$$

◻

If an algorithm can be completely decomposed into n parallelizable units without loss of efficiency then the Speedup obtained is

$$\text{Speedup}\,(n) \;=\; \frac{T_{SEQ}}{\dfrac{T_{SEQ}}{n}} \;=\; n \tag{2.48}$$

If however, only a fraction, f, of the algorithm is parallelizable then the speedup obtained is

$$\text{Speedup}\,(n) \;=\; \frac{T_{SEQ}}{\left((1-f) + \dfrac{f}{n}\right)T_{SEQ}} \;=\; \frac{1}{1-f+\dfrac{f}{n}} \tag{2.49}$$

which yields

$$\lim_{n \to \infty} \left(\text{Speedup}\,(n)\right) \;=\; \frac{1}{1-f} \tag{2.50}$$

This is known as Amdahl's Law. The ratio shows that even with an infinite amount of computing power an algorithm with a sequential component can only achieve the speedup in Eq. 2.50. If an algorithm is 50% sequential then the maximum speedup achievable is 2. While this may be a strong argument against the merits of parallel processing there are many important problems which have almost no sequential components.

Definition 2.22

The efficiency of an algorithm executing on n processors is defined as the ratio of the speedup to the number of processors:

$$\text{Efficiency} (n) = \frac{\text{Speedup} (n)}{n} \tag{2.51}$$

❏

Using Amdahl's law

$$\text{Efficiency} (n) = \frac{1}{n (1 - f) + f} \tag{2.52}$$

with

$$\lim_{n \to \infty} (\text{Efficiency} (n)) = 0 \qquad \text{when } f \neq 1 \tag{2.53}$$

2.5.2 Pipelining

Pipelining is a means to achieve speedup for an algorithm by dividing the algorithm into stages. Each stage is to be executed in the same amount of time. The flow is divided into k distinct stages. The output of the jth stage becomes the input to the $(j + 1)$ th stage. Pipelining is illustrated in Figure 2.13. As

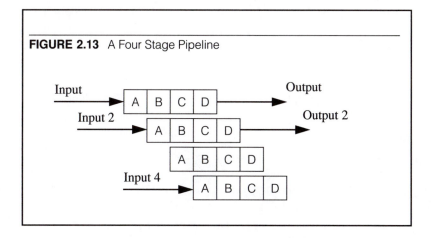

FIGURE 2.13 A Four Stage Pipeline

seen in the figure the first output is ready after four time steps Each subsequent output is ready after one additional time step. Pipelining becomes efficient when more than one output is required. For many algorithms it may not be possible to subdivide the task into k equal stages to create the pipeline. When this is the case a performance hit will be taken in generating the first output as illustrated in Figure 2.14.

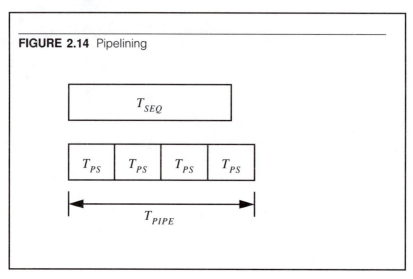

FIGURE 2.14 Pipelining

In the figure T_{SEQ} is the time for the algorithm to execute sequentially. T_{PS} is the time for each pipeline stage to execute. T_{PIPE} is the time to flow through the pipe. The calculation of the time complexity sequence to process n inputs yields

$$T_{SEQ}(n) = nT_{SEQ} \tag{2.54}$$

$$T_{PIPE}(n) = kT_{PS} + (n-1)T_{PS} \tag{2.55}$$

for a k-stage pipe. It follows that $T_{PIPE}(n) < T_{SEQ}(n)$ when

$$n > \frac{T_{PS}(k-1)}{T_{SEQ} - T_{PS}} \tag{2.56}$$

The speedup for pipelining is

$$S(n) = \frac{T_{SEQ}(n)}{T_{PIPE}(n)} = \frac{T_{SEQ}}{T_{PS} - \frac{(k-1)T_{PS}}{n}} \tag{2.57}$$

EXAMPLE 2.6 Order

$$f_1 = \left\{ \begin{array}{ll} 1, & n < 10^{50} \\ e^n, & n \geq 10^{50} \end{array} \right.$$

$$f_2 = \left\{ \begin{array}{ll} e^n, & n < 10^{50} \\ 1, & n \geq 10^{50} \end{array} \right.$$

$$f_1 \in \Theta(e^n) \text{ and } f_2 \in \Theta(1)$$

---

which yields

$$\lim_{n \to \infty} S(n) = \frac{T_{SEQ}}{T_{PS}} \qquad (2.58)$$

In some applications it may not be possible to keep the pipeline full at all times. This can occur when there are dependencies on the output. This is illustrated in Example 2.7. For this case let us assume that the addition/subtraction

---

EXAMPLE 2.7 Output Dependency PseudoCode

1 If $x > 3$ then

2      $y = y + 4;$

    else

3      $y = y - 2;$

---

operation has been set up as a pipeline. The first statement in the pseudo-code will cause the inputs x and 3 to be input to the pipeline for subtraction. After the first stage of the pipeline is complete, however, the next operation is unknown. In this case, the result of the first statement must be established. To determine the next operation the first operation must be allowed to proceed through the pipe. After its completion the next operation will be determined. This process is referred to flushing the pipe. The speedup obtained with flushing is demonstrated in Example 2.8.

EXAMPLE 2.8   Pipelining

Determine the speedup, in the limit, for a k-stage pipe over a sequential algorithm if the pipe has to be flushed 40% of the time.

Solution:

$$T_{PIPE}(n) = (0.4n) kT_{PS} + (0.6n) T_{PS}$$

$$S(n) = \frac{T_{SEQ}(n)}{T_{PIPE}(n)} = \frac{T_{SEQ}}{T_{PS}(0.4k + 0.6)}$$

2.5.3   Parallel Processing and Processor Topologies

There are a number of common topologies used in parallel processing. Algorithms are increasingly being developed for the parallel processing environment. Many of these topologies are widely used and have been studied in great detail. The topologies presented here are

- Full Crossbar
- Rectangular Mesh
- Hypercube
- Cube-Connected Cycles

2.5.3.1   Full Crossbar

A full crossbar topology provides connections between any two processors. This is the most complex connection topology and requires $(n(n-1))/2$ connections. A full crossbar is shown in Figure 2.15.

In the graphical representation the crossbar has the set, V, and E with

---

FIGURE 2.15 Full Crossbar Topology

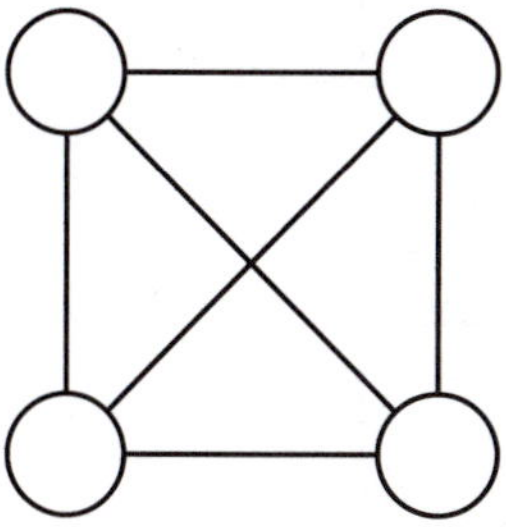

Each processor has a connection to every other processor.

$$V = \{p_i, 0 \leq i < n\} \qquad (2.59)$$

$$E = \{(p_i, p_j), 0 \leq i < n, 0 \leq j < n\} \qquad (2.60)$$

Because of the large number of edges the topology is impractical in design for large n.

2.5.3.2 Rectangular Mesh

A rectangular mesh topology is illustrated in Figure 2.16. From an implementation aspect the topology is easily scalable. The degree of each node in a rectangular mesh is at most four. A processor on the interior of the mesh has neighbors to the north, east, south, and west. There are several ways to implement the exterior nodes if it is desired to maintain that all nodes have the same degree. For an example of the external edge connection see Problem 2.5.

2.5.3.3 Hypercube

A hypercube topology is shown in Figure 2.17. If the number of nodes, n, in the hypercube satisfies $n = 2^d$ then the degree of each node is d or $\log(n)$. As a result, as n becomes large the number of edges of each node increases. The magnitude of the increase is clearly more manageable than that of the full crossbar but it can still be a significant problem with hypercube architectures containing 64K nodes. As a result the cube-connected cycles, described in the next section, becomes more attractive due to its fixed degree.

The vertices of an n dimensional hypercube are readily described by the binary ordered pair

$$(x_0, x_2, \ldots, x_{d-1}) \qquad x_j \in \{0, 1\} \qquad (2.61)$$

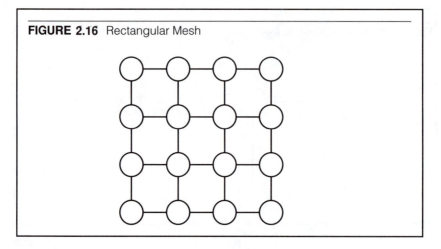

FIGURE 2.16 Rectangular Mesh

With this description two nodes are neighbors if they differ in their representation in one location only. For example for an 8 node hypercube with nodes enumerated

$$(0, 0, 0) \qquad (0, 0, 1) \qquad (0, 1, 0) \qquad (0, 1, 1)$$
$$(1, 0, 0) \qquad (1, 0, 1) \qquad (1, 1, 0) \qquad (1, 1, 1)$$

(2.62)

processor $(0, 1, 0)$ has three neighbors:

$$(0, 1, 1) \qquad (0, 0, 0) \qquad (1, 1, 0)$$

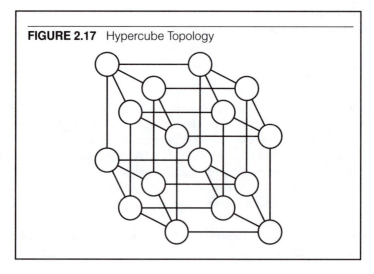

FIGURE 2.17 Hypercube Topology

2.5.3.4 Cube-Connected Cycles

A cube-connected cycles topology is shown in Figure 2.18. This topology is easily formed from the hypercube topology by replacing each hypercube node with a cycle of nodes. As a result, the new topology has nodes, each of which, has degree 3. This has the look and feel of a hypercube yet without the high degree. The cube-connected cycles topology has $n \log n$ nodes.

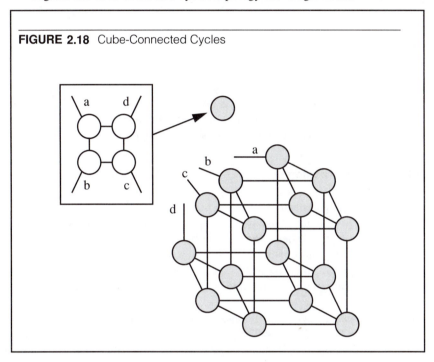

FIGURE 2.18 Cube-Connected Cycles

2.6 The Hypercube Topology

This section presents algorithms and issues related to the hypercube topology. The hypercube is important due to its flexibility to efficiently simulate topologies of a similar size.

2.6.1 Definitions

Processors in a hypercube are numbered $0, \ldots, n-1$. The dimension, d, of a hypercube, is given as

$$d = \log n \qquad\qquad (2.63)$$

where at this point it is assumed that n is a power of 2. A processor, x, in a hypercube has a representation of

$$x = (x_0, x_1, \ldots, x_{d-1}) \qquad x_j \in \{0, 1\} \qquad (2.64)$$

For a simple example of the enumeration scheme see Section 2.5.3.3 on page 75. The distance, $d(x, y)$, between two nodes x and y in a hypercube is given as

$$d(x, y) = \sum_{k=0}^{d-1} |x_k - y_k| \qquad (2.65)$$

The distance between two nodes is the length of the shortest path connecting the nodes. Two processors, x and y are neighbors if $d(x, y) = 1$. The hypercubes of dimension two and three are shown in Figure 2.19.

2.6.2 Message Passing

A common requirement of a parallel processing topology is the ability to support broadcast and message passing algorithms between processors. A broadcast operation is an operation which supports a single processor communicating information to all other processors. A message passing algorithm supports a single message transfer from one processor to the next. In all cases the messages are required to traverse the edges of the topology.

To illustrate message passing consider the case of determining the path to send a message from processor 0 to processor 7 in a 3-dimensional hypercube as shown in Figure 2.19. If the message is to traverse a path which is of minimal length, that is $d(0, 7)$, then it should travel over three edges. For this case there are six possible paths:

$$000 - 001 - 011 - 111$$

$$000 - 001 - 101 - 111$$

$$000 - 010 - 011 - 111$$

$$000 - 010 - 110 - 111$$

$$000 - 100 - 101 - 111$$

$$000 - 100 - 110 - 111$$

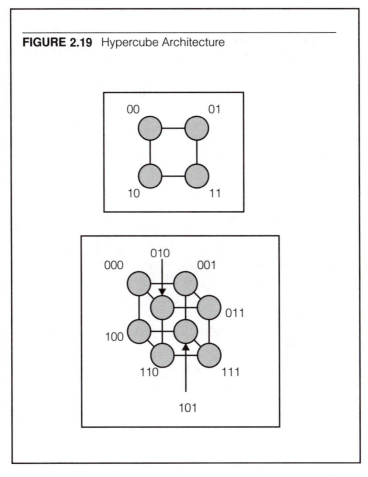

FIGURE 2.19 Hypercube Architecture

In general, in a hypercube of dimension d, a message travelling from processor x to processor y has $d(x, y)!$ distinct paths (see Problem 2.11). One simple algorithm is to compute the exclusive-or of the source and destination processors and traverse the edge corresponding to complementing the first bit that is set. This is illustrated in Table 2.4 for left to right complementing and in Table 2.5 for right to left complementing.

TABLE 2.4 Calculating the Message Path — Left to Right

Processor Source	Processor Destination	Exclusive-Or	Next Processor
000	111	111	100
100	111	011	110
110	111	001	111

TABLE 2.5 Calculating the Message Path — Right to Left

Processor Source	Processor Destination	Exclusive-Or	Next Processor
000	111	111	001
001	111	110	011
011	111	100	111

The message passing algorithm still works under certain circumstances even when the hypercube has nodes that are faulty. This is discussed in the next section.

2.6.3 Efficient Hypercubes

This section presents the analysis of the class of hypercubes for which the message passing routines of the previous section are valid. Examples are presented in detail for an 8-node hypercube.

2.6.3.1 Transitive Closure

Definition 2.23

The adjacency matrix, A, of a graph, G, is the matrix with elements a_{ij} such that $a_{ij} = 1$ implies there is an edge from i to j. If there is no edge then $a_{ij} = 0$.

❑

The adjacency matrix, A, of the transitive closure of the 8-node hypercube is simply the matrix

$$A = \begin{bmatrix} 1 & 1 & 1 & 1 & 1 & 1 & 1 & 1 \\ 1 & 1 & 1 & 1 & 1 & 1 & 1 & 1 \\ 1 & 1 & 1 & 1 & 1 & 1 & 1 & 1 \\ 1 & 1 & 1 & 1 & 1 & 1 & 1 & 1 \\ 1 & 1 & 1 & 1 & 1 & 1 & 1 & 1 \\ 1 & 1 & 1 & 1 & 1 & 1 & 1 & 1 \\ 1 & 1 & 1 & 1 & 1 & 1 & 1 & 1 \\ 1 & 1 & 1 & 1 & 1 & 1 & 1 & 1 \end{bmatrix} \tag{2.66}$$

For a hypercube with all functional nodes every processor is reachable.

2.6.3.2 Least-Weighted Path-Length

Definition 2.24

The *least-weighted path-length* graph is the directed graph where the weights of each edge correspond to the shortest path-length between the nodes.

◻

The associated weighted matrix consists of the path-length between the nodes. The path-length between a processor and itself is defined to be zero. The associated weighted matrix for an 8-node hypercube with all functional nodes is

$$A = \begin{bmatrix} 0 & 1 & 1 & 2 & 1 & 2 & 2 & 3 \\ 1 & 0 & 2 & 1 & 2 & 1 & 3 & 2 \\ 1 & 2 & 0 & 1 & 2 & 3 & 1 & 2 \\ 2 & 1 & 1 & 0 & 3 & 2 & 2 & 1 \\ 1 & 2 & 2 & 3 & 0 & 1 & 1 & 2 \\ 2 & 1 & 3 & 2 & 1 & 0 & 2 & 1 \\ 2 & 3 & 1 & 2 & 1 & 2 & 0 & 1 \\ 3 & 2 & 2 & 1 & 2 & 1 & 1 & 0 \end{bmatrix} \qquad (2.67)$$

a_{ij} is the distance between nodes i and j. If nodes i and j are not connected via any path then $a_{ij} = \infty$.

2.6.3.3 Hypercubes with Failed Nodes

This section introduces the scenario of failed processors. It is assumed if a processors or node fails then all edges incident on the processor are removed from the graph. The remaining processors will attempt to function as a working subset while still using the message passing algorithms of the previous sections. This will lead to a characterization of subcubes of a hypercube which support message passing. Consider the scenario illustrated in Figure 2.20. In the figure there are three scenarios with failed processors.

In Figure 2.20b a single processor has failed. The remaining processors can communicate with each other using a simple modification of the algorithm which traverses the first existing edge encountered.

Similarly, in Figure 2.20c communication is still supported via the modified algorithm. This is illustrated in Table 2.6. Notice that in Table 2.6 the next processor after 000 was 001. For the topology in the figure the processor did not exist so the algorithm proceeded to the next bit from right to left which gave 010. Since this processor existed the message was sent along the path.

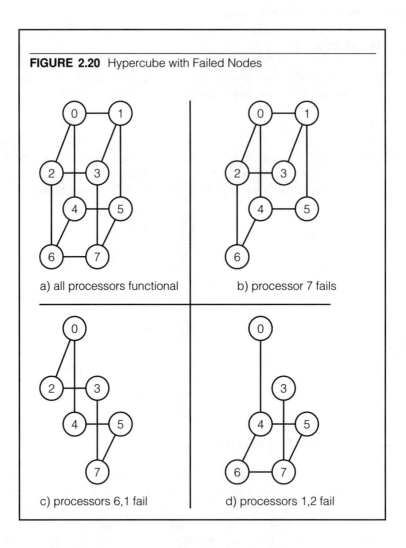

FIGURE 2.20 Hypercube with Failed Nodes

a) all processors functional

b) processor 7 fails

c) processors 6,1 fail

d) processors 1,2 fail

TABLE 2.6 Calculating the Message Path — Right to Left for Figure 2.20c

Processor Source	Processor Destination	Exclusive-Or	Next Processor
000	111	111	010
010	111	101	011
011	111	100	111

The scenario in Figure 2.20d is quite different. This is illustrated in Table 2.7.

In this case, the first processor considered to is 001 but it is not functional. Processor 010 is considered next but it is not functional. For this case the modified algorithm has failed to route the message from processor 000 to 011. There exists a path from 000 to 011 one of which is

$$000 - 100 - 101 - 111 - 011$$

Notice that the distance between the processors has increased as a result of the two processors failures. This attribute is the motivation for the characterization of efficient hypercubes in the next section.

TABLE 2.7 Calculating the Message Path — Right to Left for Figure 2.20d

Processor Source	Processor Destination	Exclusive-Or	Next Processor
000	011	011	?

2.6.3.4 Efficiency

Definition 2.25

A subcube of a hypercube is *efficient* if the distance between any two functional processors in the subcube is the same as the distance in the hypercube.

❒

A subcube with this property is referred to as an efficient hypercube. This is equivalent to saying that if A represents the least-weighted path-length matrix of the hypercube and B represents the least-weighted path-length matrix of the efficient subcube then if i and j are functional processors in the subcube then $b_{ij} = a_{ij}$. This elegant result is proven in Problem 2.20. The least-weighted path-length matrix for efficient hypercubes place ∞'s in column i and row i if processor i is failed.

The cubes in Figure 2.20b and c are efficient while the cube in Figure 2.20d is not efficient. If the cube is efficient then the modified message passing algorithm in the previous section works. The next section implements the procedure for hypercubes with failed nodes.

2.6.3.5 Message Passing in Efficient Hypercubes

The code to simulate message passing in an efficient hypercube is shown in Code List 2.8. The output of the program is shown in Code List 2.9. The path for communicating from 0 to 63 is given as 0–1–3–7–15–31–63 as shown in

Code List 2.8   Message Passing in an Efficient Hypercube

```
C++ Source Code
#include <iostream.h>


#define TRUE 1
#define FALSE 0
#define ACTIVE 1
#define INACTIVE 0
#define NO_PROCESSORS 64
#define DIMENSION 6
class node {
private:
int number;
int status;
public:
          node(int num=0) { number=0; status=ACTIVE;};
          int proc_num() { return number;};
          int pstatus() { return status;};
          void set_status(int stat) { status=stat; };
          void set_num(int num) {number = num;};
};


node hyp[NO_PROCESSORS];


void path_calc(node p1, node p2)
{
int p3;
int edge;
int i,j,z;
int ex_or;
p3=p1.proc_num();
edge =1;
cout << "Calculating path from " << p1.proc_num() <<
          " to " << p2.proc_num() << endl;
```

Code List 2.8 Message Passing in an Efficient Hypercube (continued)

```
C++ Source Code
cout << p1.proc_num() << " ";
for(j=0;j<DIMENSION;j++) {
ex_or=p3^p2.proc_num();
 edge=1;
 z=ex_or;
 for(i=0;i<DIMENSION;i++)
          {
          if((z%2==1)&&(hyp[p3^edge]).pstatus()==ACTIVE) {
          cout << hyp[p3^edge].proc_num() << " ";
          p3=p3^edge;
          break;
          }
 edge*=2;
 z=z>>1;
 }
}
cout << " Inactive Processors: ";
for(j=0; j<NO_PROCESSORS; j++) if (hyp[j].pstatus()==INACTIVE)
          cout << hyp[j].proc_num() << " ";
          cout << endl
          << "***************************************"
          << "********************"
          << endl;
};
void init_cube()
{
int i;
for(i=0;i<NO_PROCESSORS;i++) hyp[i].set_num(i);
};
void main()
{
init_cube();
```

Code List 2.8 Message Passing in an Efficient Hypercube (continued)

C++ Source Code

```cpp
path_calc(hyp[0],hyp[63]);
hyp[31].set_status(INACTIVE);
path_calc(hyp[0],hyp[63]);
hyp[15].set_status(INACTIVE);
path_calc(hyp[0],hyp[63]);
hyp[1].set_status(INACTIVE);
path_calc(hyp[0],hyp[63]);
hyp[2].set_status(INACTIVE);
path_calc(hyp[0],hyp[63]);
hyp[7].set_status(INACTIVE);
hyp[23].set_status(INACTIVE);
hyp[55].set_status(INACTIVE);
path_calc(hyp[0],hyp[63]);
path_calc(hyp[42],hyp[6]);
};
```

Code List 2.9 Output of Program in Code List 2.8

C++ Output

```
Calculating path from 0 to 63
0 1 3 7 15 31 63 Inactive Processors:
***********************************************************
Calculating path from 0 to 63
0 1 3 7 15 47 63 Inactive Processors: 31
***********************************************************
Calculating path from 0 to 63
0 1 3 7 23 55 63 Inactive Processors: 15 31
***********************************************************
Calculating path from 0 to 63
0 2 3 7 23 55 63 Inactive Processors: 1 15 31
***********************************************************
Calculating path from 0 to 63
```

Code List 2.9   Output of Program in Code List 2.8 (continued)

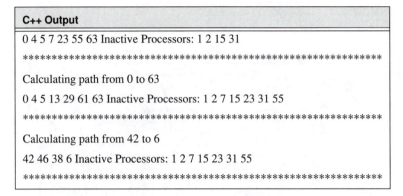

C++ Output
0 4 5 7 23 55 63 Inactive Processors: 1 2 15 31
****
Calculating path from 0 to 63
0 4 5 13 29 61 63 Inactive Processors: 1 2 7 15 23 31 55
****
Calculating path from 42 to 6
42 46 38 6 Inactive Processors: 1 2 7 15 23 31 55
****

Code List 2.9. Subsequently processor 31 is deactivated and a new path is calculated as 0–1–3–7–15–47–63 which avoids processor 31 and traverses
remaining edges in the cube. The program continues to remove nodes from
the cube and still calculates the path. All the subcubes created result in an efficient subcube.

2.6.4 Visualizing the Hypercube: A C++ Example

This section presents a C++ program to visualize the hypercube. A program to
visualize the cube is shown in Code List 2.10. The program was used to generate the PostScript image in Figure 2.21 for a 64 node hypercube. The program
uses a class structure similar to the program to visualize the Tower of Hanoi in
Code List 2.6.

The program introduces a new PostScript construct to draw and fill a circle

$$x\ y\ radius\ angle1\ angle2\ arc$$

The program uses the *scale* operator to force the image to fill a specified area.
To illustrate this, notice that the program generated both Figure 2.21 and
Figure 2.22. The nodes in Figure 2.22 are enlarged via the scale operator
while the nodes in Figure 2.21 are reduced accordingly.

The strategy in drawing the hypercube is such that only at most two processors appear in any fixed horizontal or vertical line. The cube is grown by repli-
cations to the right and downward.

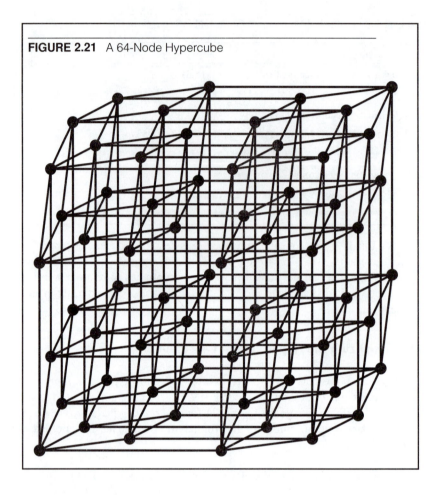

FIGURE 2.21 A 64-Node Hypercube

Code List 2.10 C++ Code to Visualize the Hypercube

C++ Code
#include <iostream.h>
#include <iomanip.h>
#include <fstream.h>
#include <math.h>
#define NO_PROCESSORS 64
#define DIMENSION 6
#define RIGHT 1

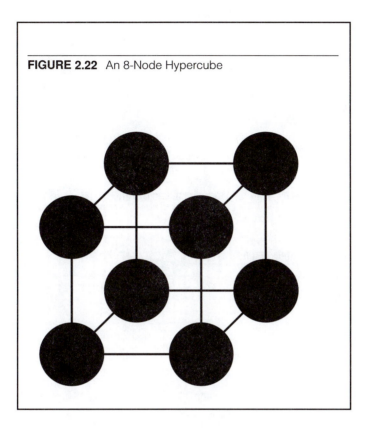

FIGURE 2.22 An 8-Node Hypercube

Code List 2.10 C++ Code to Visualize the Hypercube (continued)

C++ Code
#define DOWN 0
class cube
{
public:
int x; int y;
};
class graphics_context {
protected:
double line_w;

Code List 2.10   C++ Code to Visualize the Hypercube(continued)

```
C++ Code
                        double x_scale, y_scale;
                        double gray;
            public:
                        void set_line_width(double x=1.0) {line_w=x;};
                        void set_scale(double x=1.0, double y=1.0)
                                    {x_scale=x;y_scale=y;}
                        void set_gray(double x=0.0) {gray=x;};
};


class object : public graphics_context {
            protected:
                        double x_loc, y_loc;


            public:
                        void location(double x=0.0, double y=0.0)
                                    { x_loc=x; y_loc=y; };
                        double xlocation() { return x_loc; };
                        double ylocation() { return y_loc; };
};


class node : public object {
            protected:
            double radius;
            private:
                        int number;
            public:
            node(int num=0, double x=0.0, double y=0.0, double r=0.5)
            {           number=num;
                        x_loc=x; y_loc=y; radius=r;
                        set_scale();
                        set_line_width();
                        set_gray();
```

Code List 2.10 C++ Code to Visualize the Hypercube (continued)

C++ Code

```cpp
        };
        int proc_num() { return number; };
        void set_num(int num) {number = num;};
        void set_location(double x, double y) { x_loc=x; y_loc=y; };
        void set_radius(double r) { radius=r; };
        void draw(ofstream& file)
                {
                file << "newpath" << endl;
                file << line_w << " setlinewidth" << endl;
                file << gray << " setgray" << endl;
                file << x_loc << " " << y_loc << " " << radius <<
                        " 0 360 arc fill" << endl;
                };

        };
node hyp[NO_PROCESSORS];
int p2(int x)
        {
        return (int) pow(2.0,(double) x);
        };
void init_cube()
{
int i;
int dim,line,processor;
int translation;
int direction=RIGHT;
cube temp[NO_PROCESSORS];
for(i=0;i<NO_PROCESSORS;i++) hyp[i].set_num(i);
hyp[0].set_location(1,1);
hyp[1].set_location(2,1);
hyp[2].set_location(1,2);
hyp[3].set_location(2,2);
```

Code List 2.10   C++ Code to Visualize the Hypercube (continued)

```cpp
for(i=0;i<NO_PROCESSORS;i++)
        {
        temp[i].x=hyp[i].xlocation();
        temp[i].y=hyp[i].ylocation();
        }
for(dim=3;dim<=DIMENSION;dim++) {
        translation=1;
        for(line=1;line<=p2(dim-1);line++)
        {
                        for(processor=0;processor<p2(dim-1);processor++)
                        {
                                if(direction==RIGHT)
                                if(hyp[processor].ylocation()==line)
                                 temp[processor].y=translation++;
                                if(direction==DOWN)
                                if(hyp[processor].xlocation()==line)
                                 temp[processor].x=translation++;
                        }
        }
        if(direction==RIGHT)
        {
                        for(i=0;i<p2(dim-1);i++)
                                {
                                temp[i+p2(dim-1)].x=temp[i].x+p2(dim-2);
                                temp[i+p2(dim-1)].y=temp[i].y;
                                }
        }
        if(direction==DOWN)
        {
                        for(i=0;i<p2(dim-1);i++)
                                {
                                temp[i+p2(dim-1)].y=temp[i].y+p2(dim-2);
```

Code List 2.10 C++ Code to Visualize the Hypercube(continued)

```
C++ Code
```

```cpp
                                    temp[i+p2(dim–1)].x=temp[i].x;
                            }
                }
            direction=(direction+1)%2;
            for(i=0;i<NO_PROCESSORS;i++)
                        hyp[i].set_location(temp[i].x,temp[i].y);
    }
}


void drawline(ofstream& file, int i, int j)
{
file << hyp[i].xlocation() << " " << hyp[i].ylocation()
                    << " moveto" << endl;
file << hyp[j].xlocation() << " " << hyp[j].ylocation()
                    << " lineto stroke" << endl;
}


void render_cube(ofstream& file)
{
int i;
for(i=0;i<p2(DIMENSION);i++)
    {
    int k=1,j;
    for(j=0;j<DIMENSION;j++){
            drawline(file,i,i^k);
            k*=2;
            }
    }
            for(i=0;i<NO_PROCESSORS;i++)
                    hyp[i].draw(file);
}
void main()
```

Code List 2.10 C++ Code to Visualize the Hypercube(continued)

```
C++ Code

{
            init_cube();
            ofstream file("hyper1.ps",ios::outlios::trunc);
            if(!file)
                      {
                      cout << "Could not open file\n";
                      return;
                      }
            // Add standard postscript header
            file << "%!PS–Adobe–2.0 EPSF–2.0" << endl;
            file << "%%BoundingBox: 0 0 300 300" << endl;
            file << "%%Creator: Alan Parker" << endl;
            file << "%%EndComments" << endl;
            file << "0.0 setgray" << endl;
            double scale = 300.0/(pow(2.0,DIMENSION–1)+2.0);
            file << scale << " " << scale << " scale" << endl;
            file << 1.5/scale << " setlinewidth" << endl;
            render_cube(file);
            file << "showpage" << endl << "%%Trailer" << endl;
            file.close();

}
```

Code List 2.11 Output of Program in Code List 2.10

```
C++ File Created

%!PS–Adobe–2.0 EPSF–2.0

%%BoundingBox: 0 0 300 300

%%Creator: Alan Parker

%%EndComments

0.0 setgray

50 50 scale

0.03 setlinewidth
```

Code List 2.11 Output of Program in Code List 2.10 (continued)

C++ File Created
1 1 moveto
2 2 lineto stroke
1 1 moveto
1 3 lineto stroke
1 1 moveto
3 1 lineto stroke
2 2 moveto
1 1 lineto stroke
2 2 moveto
2 4 lineto stroke
2 2 moveto
4 2 lineto stroke
1 3 moveto
2 4 lineto stroke
1 3 moveto
1 1 lineto stroke
1 3 moveto
3 3 lineto stroke
2 4 moveto
1 3 lineto stroke
2 4 moveto
2 2 lineto stroke
2 4 moveto
4 4 lineto stroke
3 1 moveto
4 2 lineto stroke
3 1 moveto
3 3 lineto stroke
3 1 moveto
1 1 lineto stroke
4 2 moveto
3 1 lineto stroke

Code List 2.11 Output of Program in Code List 2.10 (continued)

C++ File Created
4 2 moveto
4 4 lineto stroke
4 2 moveto
2 2 lineto stroke
3 3 moveto
4 4 lineto stroke
3 3 moveto
3 1 lineto stroke
3 3 moveto
1 3 lineto stroke
4 4 moveto
3 3 lineto stroke
4 4 moveto
4 2 lineto stroke
4 4 moveto
2 4 lineto stroke
newpath
1 setlinewidth
0 setgray
1 1 0.5 0 360 arc fill
newpath
1 setlinewidth
0 setgray
2 2 0.5 0 360 arc fill
newpath
1 setlinewidth
0 setgray
1 3 0.5 0 360 arc fill
newpath
1 setlinewidth
0 setgray
2 4 0.5 0 360 arc fill

Code List 2.11 Output of Program in Code List 2.10 (continued)

C++ File Created
newpath
1 setlinewidth
0 setgray
3 1 0.5 0 360 arc fill
newpath
1 setlinewidth
0 setgray
4 2 0.5 0 360 arc fill
newpath
1 setlinewidth
0 setgray
3 3 0.5 0 360 arc fill
newpath
1 setlinewidth
0 setgray
4 4 0.5 0 360 arc fill
showpage
%%Trailer

2.7 Problems

(2.1) [Infinite Descent — Difficult] Prove, using infinite descent, that there are no solutions in the positive integers to

$$x^4 + y^4 = z^4$$

(2.2) [Recurrence] Find the closed form solution to the recursion relation

$$F(0) = a$$

$$F(1) = b$$

$$F(n) = F(n-1) - F(n-2)$$

and write a C++ program to calculate the series via the closed form solution and print out the first twenty terms of the series for

$$a = 5 \qquad b = -5$$

(2.3) [Tower of Hanoi] Write a C++ Program to solve the Tower of Hanoi problem for arbitrary n. This program should output the move sequence for a specific solution.

(2.4) [Tower of Hanoi] Is the minimal solution to the Tower of Hanoi problem unique? Prove or disprove your answer.

(2.5) [Rectangular Mesh] Given an 8x8 rectangular mesh with no additional edge connections calculate the largest distance between two processors, where the distance is defined as the minimum number of edges to traverse in a path connecting the two processors.

(2.6) [Rectangular Mesh] For a rectangular mesh with no additional edge connections formally describe the topology in terms of vertices and edges.

(2.7) [Rectangular Mesh] Write a C++ program to generate a PostScript image file of the rectangular mesh for $1 \le n \le 20$ without additional external edge connections. To draw a line from the current point to (x, y) use the primitive

$$x \; y \; lineto$$

followed by

$$gsave$$
$$stroke$$
$$grestore$$

to actually draw the line. Test the output by sending the output to a PostScript printer.

(2.8) [Cube-Connected Cycles] Calculate the number of edges in a cube connected cycles topology with $n \log n$ nodes.

(2.9) [Tree Structure] For a graph G, which is a tree, prove that

$$order\,(G) \;=\; size\,(G) + 1$$

(2.10) [Cube-Connected Cycles] For a cube-connected cycles topology formally describe the topology in terms of vertices and edges.

(2.11) [Hypercube] Given two arbitrary nodes in a hypercube of dimension n calculate the number of distinct shortest paths which connect two distinct nodes, A and B, as a function of the two nodes. Use a binary representation for each of the nodes:

$$A = \{a_0, a_1, ..., a_{n-1}\} \qquad B = \{b_0, b_1, ..., b_{n-1}\}$$
$$a_i, b_i \in \{0, 1\}$$

(2.12) [Hypercube] Given a hypercube graph of dimension n and two processors A and B what is the minimum number of edges that can be removed such that there is no path from A to B.

(2.13) Is every edge in a tree a bridge?

(2.14) Devise a broadcast algorithm for a hypercube of arbitrary dimension. Write a C++ program to simulate this broadcast operation on an 8-dimensional hypercube.

(2.15) Devise a message passing algorithm for a hypercube of arbitrary dimension. Write a C++ program to simulate this algorithm and demonstrate it for a 12-dimensional hypercube.

(2.16) Write a C++ program to visualize a complete binary tree. Your program should scale the node sizes to fit on the page as a function of the dimension in a similar fashion to Code List 2.10.

(2.17) Describe in detail the function of each procedure in the code to visualize the hypercube in Code List 2.10. Present a high-level description of the procedures *render_cube* and *init_cube*.

(2.18) Write a C++ program to display the modified adjacency matrix of an n-dimensional hypercube similar to the matrix presented in Eq. 2.67.

(2.19) Write a C++ program to visualize a 64-node hypercube which supports message passing. Your program should use a separate gray level to draw the source and destination processors and should draw the edges which form the path in a different gray scale also.

(2.20) [Difficult] Prove that the modified message passing algorithm works for any two functional processors in an efficient hypercube.

(2.21) Write a C++ program to determine if a hypercube with failed nodes is efficient.

(2.22) Calculate the least-weighted path-length matrix for each of the subcubes in Figure 2.20.

(2.23) Given a hypercube of dimension d calculate the probability that a subcube is efficient where the subcube is formed by the random failure of two processors.

(2.24) Modify the C++ program in Code List 2.10 to change the line width relative to the node size. Test out the program for small and high dimensions.

(2.25) Rewrite Code List 2.10 to build the hypercube using a recursive function.

(2.26) The program in Code List 2.10 uses a simple algorithm to draw a line from each processor node to its neighbors. As a result, the edges are drawn multiple times within in the file. Rewrite the program to draw each line only once.

3 Data Structures and Searching

This chapter introduces data structures and presents algorithms for searching and sorting.

3.1 Pointers and Dynamic Memory Allocation

This section investigates pointers and dynamic memory allocation in C++. As a first example consider the C++ source code in Code List 3.1.

Code List 3.1 Integer Pointer Example

```
C++ Source Code

void main()
{
int *p, k;
p = new int;
*p=7;
k=3;
delete p;
p=&k;
*p=4;
}
```

At the beginning of the program there are two variables that are allocated. The first variable is a variable *p* which is declared as a pointer to an integer. The second variable, *k*, is declared as an integer. The variable *p* is stored at address A1. The address A1 will contain an address of a variable which will be interpreted as an integer. Initially this address is not assigned. The variable k is stored at address A3. Note that the addresses of p and k do not change during the execution of the program. These addresses are allocated initially and belong to the program for its execution life.

The statement $p=new$ int in the program allocates room for an integer in memory and makes the pointer p point to that location. It does not assign a value to the location that p points to. In this case p now contains the address A4. The memory location at address A4 will contain an integer. The *new* operator is a request for memory allocation. It returns a pointer to the memory type requested. In this example room is requested for an integer.

The statement *$*p=7$* assigns the integer 7 to the location that p points to. In this case the address A4 will now contain a 7.

The statement $k=3$ assigns 3 to the address where k is located. In this case the address A3 will contain the integer 3.

The statement *delete p* now requests to deallocate the memory granted to p with the *new* operator. In this case p will still point to the location but the data at the location is subject to change. It can be the case that *$*p$* is no longer 7. Note that once the memory is freed the program no longer may have a right to access the data. The memory location A4 is free to be assigned to any other program which requests memory space.

The statement *$p=\&k$* assigns the address of k to p. The address of k is A3. For this case, p, located at A1 will now contain the address A3.

The statement *$*p=4$* now assigns the integer 4 to the address that p points to. For this case the data at address A3 will now contain 4.

This statement has changed the value of k. The flow for the memory is shown in Figure 3.1.

There are a number of pitfalls to be concerned with pointers. The declaration *int *p* does not allocate room for the integer. It simply allocates room for a variable p which will point to an integer in memory. As a result the following code segment is invalid:

int *p;

*p=7;

For this code segment the address that p contains is not valid. Unfortunately depending on the platform you are using to develop your programs this might

FIGURE 3.1 Memory Layout for C++ Program

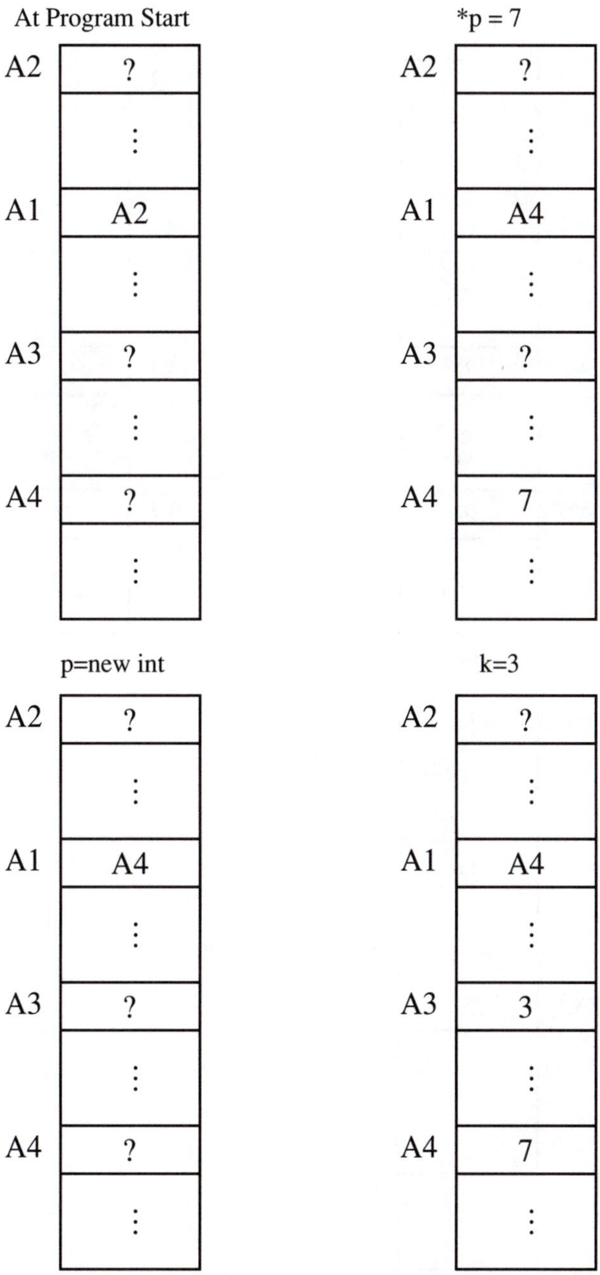

At Program Start

A2	?
	⋮
A1	A2
	⋮
A3	?
	⋮
A4	?
	⋮

*p = 7

A2	?
	⋮
A1	A4
	⋮
A3	?
	⋮
A4	7
	⋮

p=new int

A2	?
	⋮
A1	A4
	⋮
A3	?
	⋮
A4	?
	⋮

k=3

A2	?
	⋮
A1	A4
	⋮
A3	3
	⋮
A4	7
	⋮

FIGURE 3.1 Memory Layout for C++ Program (continued)

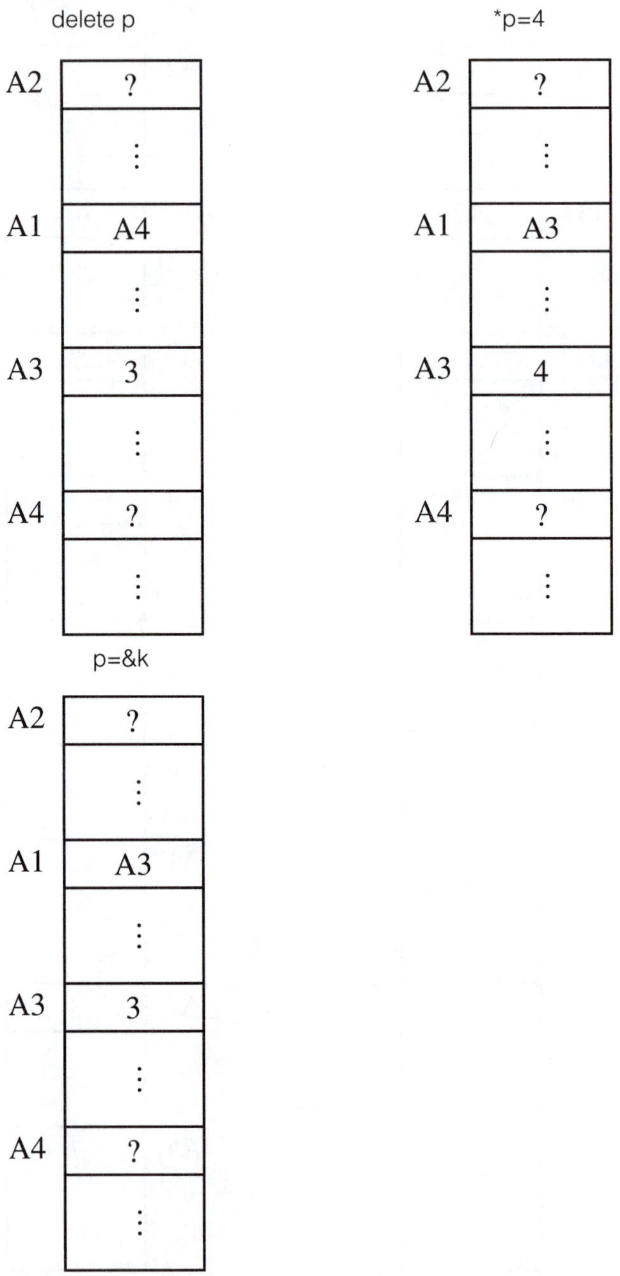

delete p

A2	?
	⋮
A1	A4
	⋮
A3	3
	⋮
A4	?
	⋮

*p=4

A2	?
	⋮
A1	A3
	⋮
A3	4
	⋮
A4	?
	⋮

p=&k

A2	?
	⋮
A1	A3
	⋮
A3	3
	⋮
A4	?
	⋮

not generate an error on compilation and in some operating systems even on execution.

The following code segment is acceptable

int *p, k;

p=&k;

*p=4;

For this code segment, *p* points to the address of k which has been allocated memory for an integer.

The code shown in Code List 3.2 is also valid. The output for the program is shown in Code List 3.3.

Code List 3.2 Pointer Example

C++ Source Code
#include <iostream.h>
void main()
{
int * * p;
p = new int *;
*p = new int;
**p=7;
cout << "The value of p is " << p << endl;
cout << "The value of *p is " << *p << endl;
cout << "The value of **p is " << **p << endl;
}

Code List 3.3 Output of Program in Code List 3.2

C++ Output
The value of p is 0x1eb4
The value of *p is 0x2530
The value of **p is 7

The style of the output will change dramatically depending on the operating system and platform used to develop the code. It is sufficient to note that for the code in Code List 3.2 *p* contains an address that points to a location that contains an address that points to a location that contains an integer.

3.1.1 A Double Pointer Example

Consider the simple program which prints out the runtime arguments provided by the user. The program source is shown in Code List 3.4. The output of the program is shown in Code List 3.5. The program is executed by typing in the command

ARGV1 arg1 arg2

Code List 3.4 Double Pointer Example

C++ Source Code

```
#include <iostream.h>
void main(int argc, char * * argv)
{
int i;
for(i=0;i<argc;i++)
cout << " Argument " << i << " is " << argv[i] << endl;
for(i=0;i<argc;i++)
cout << " Argument " << i << " is " << *(argv+i) << endl;
for(i=0;i<argc;i++,argv++)
cout << " Argument " << i << " is " << *argv << endl;
argv--;
cout << " Lets look at &((*argv)[1]) : " << &((*argv)[1]) << endl ;
cout << " Lets look at (*argv)[1] : " << (*argv)[1] << endl;
cout << " Lets look at (*argv)[4]+0x32 : " << (*argv)[4]+0x32 << endl;
cout << " Lets look at (char) (*argv)[4]+0x32 : "
          << (char) ((*argv)[4]+0x32) << endl;
// Restore argv
argv-=2;
cout << " Lets look at argv[1][1] : " << argv[1][1] << endl;
cout << " Should be the same as *(*(argv+1)+1) : " << *(*(argv+1)+1) << endl;
}
```

Code List 3.5   Output of Program in Code List 3.4

C++ Output
Argument 0 is ARGV.EXE
Argument 1 is arg1
Argument 2 is arg2
Argument 0 is ARGV.EXE
Argument 1 is arg1
Argument 2 is arg2
Argument 0 is ARGV.EXE
Argument 1 is arg1
Argument 2 is arg2
Lets look at &((*argv)[1]) : rg2
Lets look at (*argv)[1] : r
Lets look at (*argv)[4]+0x32 : 50
Lets look at (char) (*argv)[4]+0x32 : 2
Lets look at argv[1][1] : r
Should be the same as *(*(argv+1)+1) : r

The name of the program is ARGV1.EXE. The arguments passed to the program are arg1 and arg2. The main procedure receives two variables, argc and argv. For this case argc will be the integer 3 since there are 2 arguments passed to the program. It is 3 instead of 2 because argv will also hold the program name in addition to the arguments passed as can be seen in the program output. In the program argv is a pointer to a pointer to a character. The organization is shown in Figure 3.2. Looking at the figure one notes a rather complex organization. In the figure argv is stored at memory location A1. Its value is the address A2. The address A2 contains the address A5 which contains a contiguous set of characters. The first character at address A5 is the letter A (in hex 41, using ASCII). The character at address A5+1 is the letter R (in hex 52). The set of characters is terminated with a NULL character, (in hex 00). The null character indicates the end of the string. It is used by programs which are passed the address A5 to print the character. These programs print each consecutive character until they reach a NULL. A failure to place a NULL character at the end of a string will result in many string operation failures in addition to printing improperly. Remember in C/C++ a string is merely a collection of contiguous characters terminated in a NULL.

C and C++ can treat pointers as arrays. This is a very powerful and sometimes dangerous feature. For this example one can interpret

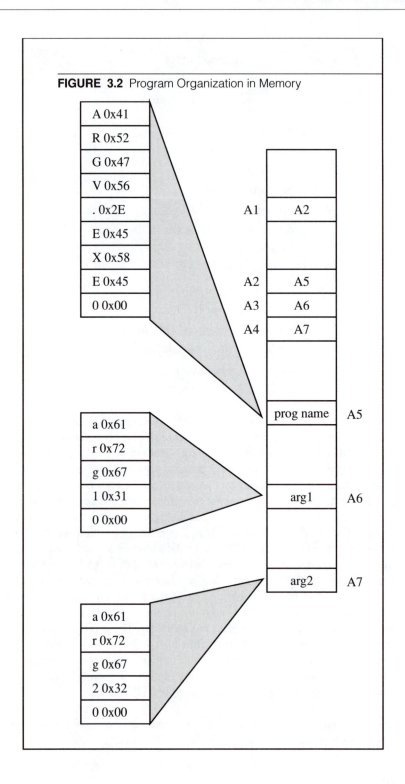

FIGURE 3.2 Program Organization in Memory

$$argv[0] = A5$$
$$argv[1] = A6$$
$$argv[2] = A7$$
$$argv[3] = undefined,$$

There are only two arguments + the program name.

Remember that argv is a pointer to a char to a char, written as char * *.

argv[0] is a char * or a pointer to a char.

When the io function cout receives a char * it will interpret the characters at the location as a string. In this case during the first print loop argv[0] points to A5 where the string representing the name of the program resides (technically, the command line argument invoking the program).

Going to the location A5 cout proceeds to print out ARGV.EXE and stops printing characters because of the NULL character reached.

C and C++ also support pointer arithmetic. This can lead to complex expressions. For this example argv+1 is synonymous with &argv[1] which in this case one has

$$argv+1 = A3$$
$$argv+2 = A4$$
$$argv+0 = A2$$
$$argv[0] = A5$$
$$argv[1] = A6$$
$$argv[2] = A7$$
$$\&argv[0] = A2$$
$$\&argv[1] = A3$$
$$\&argv[2] = A4$$
$$argv = A2$$
$$argv = \&argv[0]$$

In C and C++ when you name an array like x[10] then x with no brackets refers to the address of x[0]:

$$x = \&x[0]$$

One can traverse the pointers using * or [] that is the following is identical

$$*x = x[0]$$
$$*(x+1) = x[1]$$
$$*(x+2) = x[2]$$

Notice that

$$argv[0] = A5$$
$$argv[0][0] = `A'$$
$$argv[0][1] = `R'$$
$$argv[1][0] = `a'$$
$$argv[1][1] = `r'$$

Make sure you understand all the outputs of the program. If you are going to spend a lot of time programming in C or C++ then you should review this chapter frequently until you are completely comfortable with the concepts.

3.1.2 Dynamic Memory Allocation with New and Delete

C++ has introduced memory allocation operators *new* and *delete* to deal with requesting and freeing memory. An example of the use of *new* and *delete* are illustrated in Code List 3.6. The output of the program is shown in Code List 3.7. There are some important features of new and C++ illustrated in this program.

Code List 3.6 Dynamic Memory Allocation in C++

```
C++ Source

// This program demonstrates the differences between new
// and malloc
#include <iostream.h>
#include <malloc.h>
#include <new.h>
class
test
        {
        public:
        test() { cout << "Constructor function called" << endl;}
        ~test() { cout << "Destructor function called" << endl;}
        };
```

Code List 3.6 Dynamic Memory Allocation in C++(continued)

C++ Source
void main()
{
test * k, *j; // Declare pointers to class test
test w; // Declare a variable test to investigate constructor functions
cout << "At Point 1" << endl;
j = new test[4]; // Request an array of class objects of size 9
cout << "At Point 2" << endl;
k = (test *) malloc(4*sizeof(test)); // Request array
cout << "At Point 3" << endl;
delete[] j; // Give back memory allocated
cout << "At Point 4" << endl;
free(k); // Give back memory allocated
cout << "At Point 5" << endl;
}

Code List 3.7 Output of Program in Code List 3.6

C++ Output
Constructor function called
At Point 1
Constructor function called
Constructor function called
Constructor function called
Constructor function called
At Point 2
At Point 3
Destructor function called
Destructor function called
Destructor function called
Destructor function called
At Point 4

Code List 3.7 Output of Program in Code List 3.6 (continued)

C++ Output
At Point 5
Destructor function called

The program declares a class called *test*. Two variables *k* and *j* are declared as pointers to objects of type *test*. Upon declaration room is stored in memory for the pointers *k* and *j*.

A variable *w* of type *test* is created with the statement *test w;*. This statement illustrates the use of constructor functions in C++. When *w* is created the constructor function *test( )* is called which results in "Constructor function called" being printed.

The statement *j=new test[4];* requests memory for an array of size four for the class *test*. As a result of using *new* the constructor function is called four times. After the statement *j* will point to the first element.

The statement *k = (test *) malloc(4*sizeof(test));* requests memory for an array of size 4 for the class *test*. Using *malloc*, however, will not call the constructor function for the class *k*. As a result nothing is printed at this point of the program.

The statement *delete[] j;* gives back the memory requested by the *new* operator earlier. The brackets *[]* are used when *new* is used to declare an array. At this point the destructor function *~test( )* is called for each element in the array.

The statement *free(k)* gives back the memory allocated by the *malloc* request. As with *malloc, free* will not call the destructor function.

Before the program terminates the variable local to main *w* will first lose its scope and as a result the destructor function will be called for *w*.

In C++ *new* and *delete* should be used in lieu of *malloc* and *free* to ensure the proper calling of constructor and destructor functions for the classes allocated. Notice that *new* also avoids the use of the *sizeof* operator which simplifies its use.

3.1.3 Arrays

Sequential arrays stored in memory also rely on pointers for index calculations. The array example in Code List 3.8 demonstrates the differences between pointers and arrays for the case of the multidimensional array. The output of the program is shown for two different platforms. Code List 3.9 shows the output of the program for a DOS system while Code List 3.10 shows the output of the program on a Unix system. For this program two different methodologies are used for implementing the storage of four integers. The memory allocation is illustrated in Figure 3.3. The key difference between the implementation of the pointers and the multidimensional array is that the array *a[2][2]* is not a variable. As a result, operations such as *a=a+1* are invalid.

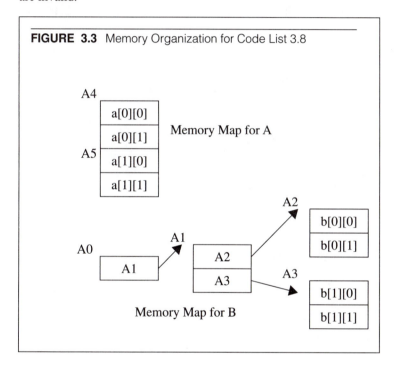

FIGURE 3.3 Memory Organization for Code List 3.8

Someone slightly familiar with C or C++ might be surprised to see that the output indicates that the values of *&a*, *a*, and **a* are all equal. While this looks unusual it is correct. The declaration *int a[2][2]* in C and C++ declares *a* to be an array of arrays. In this case there are two arrays each containing two integers. The first array is located at address A4 while the second array is located at the address A5.

- *a* - returns the starting address of the array of arrays which is given as A4 in Figure 3.3.

- **a* - returns the starting address of the first array in the list which is also A4 in Figure 3.3

- &*a* - returns the starting address of the array a which is A4. This does not return the address of the element (if there is one) that actually points to *a*. When you declare an array via int *a[2][2]* there is no variable which points to the beginning of the array that the programmer can change. The compiler basically ignores the ampersand when the variable is declared as an array. Remember, this is the difference between pointers and arrays. The location where *a* points to cannot change during the program.

The output for *b* follows directly the addressing as illustrated in Figure 3.3

Code List 3.8 Array Example

C++ Source Code
| ```
#include <iostream.h>
// This program demonstrates multidimensional addressing
// in C and C++
void main()
{
int a[2][2];
int * * b;
b = new int * [2];
b[0] = new int [2];
b[1] = new int [2];
b[0][0]=1;
b[0][1]=2;
b[1][0]=3;
b[1][1]=4;
a[0][0]=1;
a[0][1]=2;
a[1][0]=3;
a[1][1]=4;
cout << "The size of int is " << sizeof(int) << endl;
cout << "The size of a is " << sizeof(a) << endl;
cout << "The size of b is " << sizeof(b) << endl;
cout << "The value of a is " << a << endl;
cout << "The value of *(a) is " << *a << endl;
cout << "The value of &a is " << &a << endl;
``` |
```

Code List 3.8 Array Example (continued)

C++ Source Code
cout << "The value of **a is " << **a << endl; cout << "The value of a+1 is " << a+1 << endl; cout << "The value of *(a+1) is " << *(a+1) << endl; cout << "The value of **(a+1) is " << **(a+1) << endl; cout << "The value of *a[1] is " << *a[1] << endl; cout << "The value of (*a)[1] is " << (*a)[1] << endl; cout << "The value of b is " << b << endl; cout << "The value of &b is " << &b << endl; cout << "The value of b+1 is " << b+1 << endl; cout << "The value of *(b) is " << *b << endl; cout << "The value of *(b+1) is " << *(b+1) << endl; cout << "The value of **(b+1) is " << **(b+1) << endl; cout << "The value of **b is " << **b << endl; cout << "The value of (*b)[1] is " << (*b)[1] << endl; cout << "The value of *b[1] is " << *b[1] << endl; cout << "The value of b[1][0] is " << b[1][0] << endl; }

Code List 3.9 Output of Code in Code List 3.8

C++ Output (DOS)
The size of int is 2
The size of a is 8
The size of b is 2
The value of a is 0xffee
The value of *(a) is 0xffee
The value of &a is 0xffee
The value of **a is 1
The value of a+1 is 0xfff2
The value of *(a+1) is 0xfff2
The value of **(a+1) is 3
The value of *a[1] is 3

Code List 3.9 Output of Code in Code List 3.8 (continued)

C++ Output (DOS)
The value of (*a)[1] is 2
The value of b is 0x10f8
The value of &b is 0xffec
The value of b+1 is 0x10fa
The value of *(b) is 0x1100
The value of *(b+1) is 0x1108
The value of **(b+1) is 3
The value of **b is 1
The value of (*b)[1] is 2
The value of *b[1] is 3
The value of b[1][0] is 3

Code List 3.10 Output of Code in Code List 3.8

C++ Output (UNIX)
The size of int is 4
The size of a is 16
The size of b is 4
The value of a is 0xf7fffb80
The value of *(a) is 0xf7fffb80
The value of &a is 0xf7fffb80
The value of **a is 1
The value of a+1 is 0xf7fffb88
The value of *(a+1) is 0xf7fffb88
The value of **(a+1) is 3
The value of *a[1] is 3
The value of (*a)[1] is 2
The value of b is 0x1cba0
The value of &b is 0xf7fffb7c
The value of b+1 is 0x1cba4
The value of *(b) is 0x1cbb0
The value of *(b+1) is 0x1cbc0

Code List 3.10 Output of Code in Code List 3.8 (continued)

C++ Output (UNIX)
The value of **(b+1) is 3
The value of **b is 1
The value of (*b)[1] is 2
The value of *b[1] is 3
The value of b[1][0] is 3

3.1.4 Overloading in C++

An example of overloading in C++ is shown in Code List 3.11. The output of the program is shown in Code List 3.12. This program overloads the operator *()* which is used to index into a set of characters for a specific data bit. The packing is illustrated in Figure 3.4 for the variable e declared in the program.

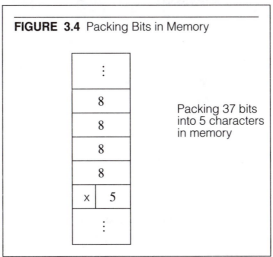

FIGURE 3.4 Packing Bits in Memory

Packing 37 bits into 5 characters in memory

Code List 3.11 Operator Overloading Example

C++ Source
// This program demonstrates packing bits in memory
// It illustrates the use of operator overloading in C++
#include <iostream.h>
class binary_data
{

Code List 3.11   Operator Overloading Example (continued)

C++ Source

```
            unsigned char * data;
            public:
            int size;
            binary_data(int size)
                    {
                    data = new unsigned char[size/8+(size%8?1:0)];
                    binary_data::size=size;
                    int i;
                     for(i=0;i<size;i++) assign(i,0);
                    }
            ~binary_data()
                    {
                    delete[] data;
                    }
            int operator() (int index);
            void assign(int index,int value);
            void print();
};
void binary_data::print()
        {
        int i;
        for(i=size−1;i>=0;i--) cout << (*this)(i);
        }
int binary_data::operator() (int index)
        {
        unsigned char mask=0x1<<index%8;
        return ((((this->data)[index/8])&mask)?1:0);
        }
void binary_data::assign(int index, int value)
        {
        if(value) data[index/8]|=value<<index%8;
        else data[index/8]&=~(0x1<<index%8);
```

Code List 3.11 Operator Overloading Example (continued)

```
C++ Source
            }
void main()
{
binary_data q(4);
binary_data d(9);
binary_data e(37);
q.assign(0,1);
q.assign(2,1);
q.assign(3,1);
q.assign(2,q(1));
d.assign(3,1);
d.assign(4,1);
e.assign(36,1);
e.assign(14,1);
cout << "The value of q is "; q.print(); cout << endl;
cout << "The value of d is "; d.print(); cout << endl;
cout << "The value of e is "; e.print(); cout << endl;
}
```

Code List 3.12 Output of Program in Code List 3.11

```
C++ Output
The value of q is 1001
The value of d is 000011000
The value of e is 1000000000000000000000100000000000000
```

3.2 Arrays

This section demonstrates the creation of an array class in C++ using templates. The goal of the program is to demonstrate the implementation of a feature of C++ which is already built in; therefore, the code is for instructive purposes only. The code for a program to create an array class is illustrated in Code List 3.13, The output of the program is shown in Code List 3.14. The array class is declared in the program as a generic class with a type T which is specified later when an array variable is declared. As seen in the main function three arrays are declared: *a*, *b*, and *c*. The array *a* consists of ten integers. The array *b* consists of five doubles. The array *c* consists of 3 characters. The constructor function for the array initializes all the elements of the array to zero. The function *set_data* is used to assign a value to a specific element in the array. The function *print_data* is used to print a specific element in the array.

Code List 3.13 Creating an Array Class in C++

C++ Source

```
// This program creates a template to create an array.
// C++ supports arrays already so this is for instructive
// purposes only
#include <iostream.h>
template<class T, int size>
class array {
private:
T data[size];
public:
        array(void);
        T get_data(int i);
        void set_data(int i, T x);
        void print_data(char * x, int i);
        };
// Initialization constructor for array
template<class T, int size>
array<T, size>::array(void)
{ int i; for(i=0;i<size;i++) data[i]= 0; }
// function to retrieve element i
template<class T, int size>
T array<T,size>::get_data(int i)
        {
        return data[i];
        }


// function to print element i
template<class T, int size>
```

Code List 3.13 Creating an Array Class in C++ (continued)

```
C++ Source
void array<T,size>::print_data(char * x, int i)

        {

        cout << x << "[" << i << "] = " << data[i] << endl;

        }
// function to assign a value to element i
template<class T, int size>
void array<T,size>::set_data(int i, T x)

        {

        data[i]=x;

        }


void main()
{

        array<int,10> a;

        array<double,5> b;

        array<char,3> c;

        a.print_data("a",3);

        b.print_data("b",4);

        b.set_data(4,4.7);

        a.set_data(3,10.8);

        a.print_data("a",3);

        b.print_data("b",4);

        c.set_data(2,'n');

        c.print_data("c",2);

}
```

Code List 3.14 Output from Code List 3.13

```
C++ output
a[3] = 0
b[4] = 0
a[3] = 10
```

Code List 3.14 Output from Code List 3.13 (continued)

C++ output
b[4] = 4.7
c[2] = n

3.3 Stacks

A stack is a data structure used to store and retrieve data. The stack supports two operations *push* and *pop*. The *push* operation places data on the stack and the *pop* operation retrieves the data from the stack. The order in which data is retrieved from the stack determines the classification of the stack.

A FIFO (First In First Out) stack retrieves data placed on the stack first. A LIFO (Last In First Out) stack retrieves data placed on the stack last. A LIFO stack *push* and *pop* operation is illustrated in Figure 3.5.

The source code to implement a LIFO stack class is shown in Code List 3.15. The output of the program is shown in Code List 3.16. Notice that templates are used again so the type used for the stack is defined at a later point.

Code List 3.15 LIFO Stack Class

```
C++ Source

// This programs creates a stack class with push and pop operations
#include <iostream.h>
//Define the stack class, set default stack size to 2
// use a template so you can define the type at a later point
template<class T,int size=2>
class stack {
        private: T data[size];
        int stack_ptr;
        public:
        stack(void);
        void push(T x);
        T pop();
        };
// Constructor function to zero elements in stack
// and to initialize data
```

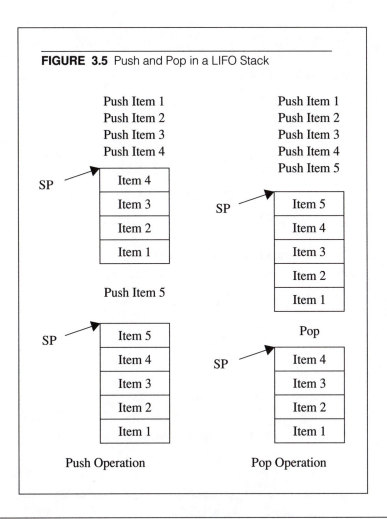

FIGURE 3.5 Push and Pop in a LIFO Stack

Code List 3.15 LIFO Stack Class (continued)

C++ Source

```
template<class T, int size>
stack<T,size>::stack(void)
        {
        int i;
        for(i=0;i<size;i++) data[i]=0;
        stack_ptr=0;
```

Code List 3.15 LIFO Stack Class (continued)

C++ Source

```cpp
        }
// push data onto stack
template<class T, int size>
void stack<T,size>::push(T x)
        {
        if(stack_ptr>=size)
                {
                cout << "Cannot push data: stack full" << endl;
                return;
                }
        data[stack_ptr++]=x;
        cout << "Placed " << x << " on stack" << endl;
        return;
        }
template<class T, int size>
T stack<T,size>::pop()
        {
        if(stack_ptr<=0)
                {
                cout << "Cannot pop data: stack empty" << endl;
                return data[0] ;
                }
        cout << "Popped " << data[--stack_ptr] << " from stack" << endl;
        return data[stack_ptr];
        }
void main()
        {
// create a stack of integers
        stack<int,10> s;
        s.push(45);
        s.pop();
// try to pop an empty stack
```

Code List 3.15 LIFO Stack Class (continued)

C++ Source

```cpp
        s.pop();
        s.push(56);
        s.push(29);
        s.push(31);
        s.pop();
        s.pop();
// create a stack of doubles
        stack<double,2> d;
        d.push(4.5);
        d.push(5.9);
// try to push on a full stack
        d.push(7.2);
        d.pop();
        d.pop();
// try to pop an empty stack
        d.pop();
// declare a stack of characters - use default size
        stack<char> c;
        char w;
        c.push('n');
        c.push('l');
// try to push on a full stack
        c.push('w');
        c.pop();
// grab the stack value
        w=c.pop();
        cout << "I got that character ** " << w <<
                " ** that was popped." << endl;
// try to pop an empty stack
        c.pop();
        }
```

Code List 3.16 Output of Program in Code List 3.15

C++ Output
Placed 45 on stack
Popped 45 from stack
Cannot pop data: stack empty
Placed 56 on stack
Placed 29 on stack
Placed 31 on stack
Popped 31 from stack
Popped 29 from stack
Placed 4.5 on stack
Placed 5.9 on stack
Cannot push data: stack full
Popped 5.9 from stack
Popped 4.5 from stack
Cannot pop data: stack empty
Placed n on stack
Placed l on stack
Cannot push data: stack full
Popped l from stack
Popped n from stack
I got that character ** n ** that was popped.
Cannot pop data: stack empty

3.4 Linked Lists

This section presents the linked list data structures. This is one of the most common structures in program design.

3.4.1 Singly Linked Lists

A linked list with four entries is shown in Figure 3.6. As seen in the figure, there is a pointer which points to the head of the list. Each object in the list has associated data and a pointer to the next element in the list. The figure is shown with four objects. The final element contains a NULL pointer. This is common practice to indicate the end of the list. The data in the linked list can be a single element or a large collection of data.

A C++ program to demonstrate the linked list is shown in Code List 3.17. This program creates a linked lists of classes. The class template is declared as

```
template <class T>
class list {
private:
        list <T> * next;
        friend class start_list<T>;
        friend class iterator<T>;
public:
        T data;
};
```

In this declaration *next* is declared as a pointer to the next element in the list. Two classes are declared as *friends* to the class, *start_list* and *iterator*. As a result these classes will have access to the functions and data of the class list. *data* is declared as public in the class. The data type T is declared later in the program.

The next class declared in the program is *start_list* which is defined as

```
class start_list
        {
        list<T> *start;
        friend class iterator<T>;
        public:
        start_list(void) { start=0;}
        ~start_list(void);
        void add(T t);
        int isMember(T t);
        }
```

For this class, a pointer to the start of a list is declared. The constructor function *start_list()* initializes start to zero when an item of class *start_list* is declared. The function *start_list()* is declared inline. The function *add* is used to add elements to the list. The destructor function *~start_list()* is called when data of type *start_list* lose their scope. The function *~start_list()* is not declared inline. The function *isMember* is used to determine if a data element matches an element in any of the members of the linked list. Notice that in the program, *start_list* is used to instantiate a class of type *list*. The *add* function is declared next in the program This function creates an element of type *list* and appends it to the current list. If the list is empty then the function assigns *start* to the beginning of the new list.

The *isMember* function is declared next in the program. The *isMember* function searches the list and tries to find a match to the data t that is passed. If a match is found the function returns 1 else the function returns 0.

The destructor function for the class, *~start_list*, is defined next. The destructor function begins at the start of the list and deletes the lists that are formed making up the entire linked list. The destructor function in turn assigns *start* to null. This function will be called in the program when any data of type *start_list* loses scope. This is a very powerful technique of C++. Typically the constructor functions are used to acquire memory upon the creation of a variable and the memory is freed up via the destructor function.

The next class defined is the *iterator* class. The *iterator* class is used to traverse the linked list. The iterator class contains a pointer to the start of a list and a *cursor* to traverse the list. The class contains a function *reset* which sets the *cursor* back to the start of the list. The constructor function for the class accepts a parameter which is a pointer to a class of type *start_list*. The constructor function calls *reset* to initialize *cursor*. The function *next* is used to iterate the list. The function assigns the pointer *p* to *cursor* and *cursor* to *cursor->next* if *cursor* is not null.

The program then initiates a number of *typedefs* which create lists and pointers to list for the data types of string, double, int, char.

The *main()* routine creates a number of lists. The first list created, *number*, is declared with *list_double number*. This list will contain a list of data elements of type double. Upon the declaration of *list_double* room for the data has not been allocated and the list pointers have been set to null. The first time room for data is allocated is during the call *number.add(4.5)*. This adds 4.5 to the list. Subsequent calls to *number.add()* append the data to the list. To access the numbers in the newly formed list a *list_double_iterator* is declared with *list_double_iterator x(&number)*. The *list_double_ptr p* access the data via calls to the iterator function *x.next()*. The output for the program is shown in Code List 3.18.

Code List 3.17 Linked List Source

C++ Linked List Source Code
// This program creates a template to create a linked list
// of data. A function to add data is provided as well as a

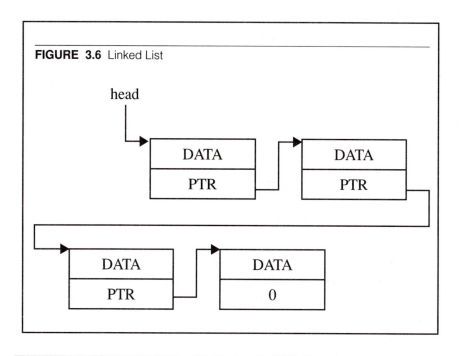

FIGURE 3.6 Linked List

Code List 3.17 Linked List Source (continued)

```
C++ Linked List Source Code

// function to search for a member


#include <iostream.h>
// This is used to instantiate the data
template<class T>
class list{
        private:
        list<T> * next;
        friend class start_list<T>;
        friend class iterator<T>;
        public:
        T data;
        };
template<class T>
```

Code List 3.17   Linked List Source (continued)

```
C++ Linked List Source Code

class start_list
        {
        list<T> * start;
        friend class iterator<T>;
        public:
        start_list(void) { start=0;}
        ~start_list(void);
        void add(T t);
         int isMember(T t);
        }


template<class T>
void start_list<T>::add(T t)
        {
        list<T> *p=start, *q=start, *r;
        while(p!=0) {q=p; p=p->next; }
        r = new list<T>;
        r->data = t; r->next = 0;
        if(start) q->next = r; else start=r;
        }


// This will not test if the list contains a specific string
template<class T>
int start_list<T>::isMember(T t)
        {
        list<T> *p=start, *q=start;
        while(p!=0)
                {
                q=p;
                p=p->next;
                 if(q->data==t) return 1;
                }
```

Code List 3.17 Linked List Source (continued)

C++ Linked List Source Code

```
            return 0;
            }

template<class T> start_list<T>::~start_list(void)
        {
        list<T> * p = start, *q;
        while(p!=0) { q=p; p=p->next; delete q; }
        start=0;
        }

template<class T>
        class iterator
        {
        start_list<T> *l;
        list<T> *cursor;
        public:
                    void reset(void) {cursor=l->start;}
                    iterator(start_list<T> *li)
                            { l=li; reset(); }
            list<T> *next(void);
            }

template<class T> list<T> * iterator<T>::next(void)
        {
        list<T> * p = cursor;
        if(cursor) cursor=cursor->next;
        return p;
        }

typedef start_list<char *> list_string;
typedef start_list<double> list_double;
typedef start_list<int> list_int;
```

Code List 3.17 Linked List Source (continued)

C++ Linked List Source Code

```cpp
typedef start_list<char> list_char;
typedef list<char *> * list_string_ptr;
typedef list<double> * list_double_ptr;
typedef list<int> * list_int_ptr;
typedef list<char> * list_char_ptr;
typedef iterator<char *> list_string_iterator;
typedef iterator<double> list_double_iterator;
typedef iterator<int> list_int_iterator;
typedef iterator<char> list_char_iterator;
void main()
        {
        list_double number;
        list_double_ptr p;
        list_string str;
        list_string_ptr q;
        number.add(4.5);
        number.add(5.7);
        number.add(3.4);
        str.add("Hello\n");
        str.add("This is a ");
        str.add("Test\n");
        list_double_iterator x(&number);
        list_string_iterator y(&str);
        cout << "List: " <<endl;
        while((p=x.next())!=0) cout << "Item " << p->data << endl;
        if(number.isMember(4.5)) cout << "4.5 is in list"<<endl;
                else cout << "4.5 is not in list"<< endl;
        if(number.isMember(4.4999))
                cout << "4.4999 is in list"<<endl;
                else cout << "4.4999 is not in list" << endl;
        cout << endl << "List: " << endl;
```

Code List 3.17 Linked List Source (continued)

C++ Linked List Source Code
while((q=y.next())!=0) cout << q->data;
};

Code List 3.18 Output from Code List 3.17

C++ Output
List:
Item 4.5
Item 5.7
Item 3.4
4.5 is in list
4.4999 is not in list
List:
Hello
This is a Test

3.4.2 Circular Lists

A circular list with two entries is shown in Figure 3.7. A circular list contains a pointer from the last object in the list to the first. In a sense, the new list has no beginning or end. The circular list is common in use for storing the most recent data when limited to finite storage. A common technique is to allocate a fixed amount of storage for a particular database and after it fills up to write over the old data by looping back around to the beginning. Obviously, the application is limited to cases where data loss is not critical. An example might be a database used to store the last 20 issues of The Wall Street Journal.

3.4.3 Doubly Linked Lists

A doubly linked list with two elements is shown in Figure 3.8. Doubly linked lists are used to provide bidirectional access to the data in the list. For many searching techniques it might be useful to traverse data from both sides of the list. A good example of this is quicksort which is discussed in Section 3.8.

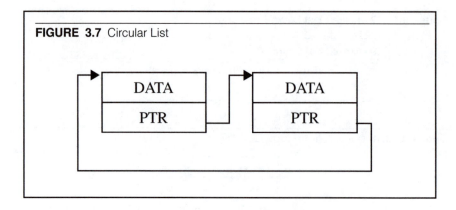

FIGURE 3.7 Circular List

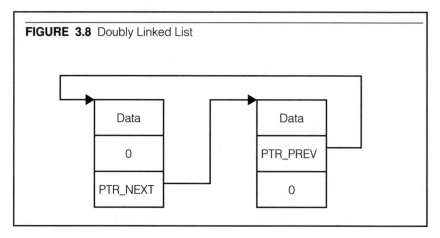

FIGURE 3.8 Doubly Linked List

3.5 Operations on Linked Lists

There are a number of operations on linked lists that are useful. These operations might be assigned to a class from which different types of linked lists are derived. Some common operations might be

- add_object — to add an object to the linked list
- destroy_object — to destroy an object of the linked list
- find_object — to find an object in the list
- find_member — to search the whole list for a specific member
- find_last_member — finds the last object in the list which matches the specific member

A number operations including sorting might also be defined for the linked list.

3.5.1 A Linked List Example

This section presents a complete example in C++ which demonstrates the use of linked lists to search for the solution to a particular coffee-house game. The purpose of the game is to eliminate as many pegs as possible on a triangular board by jumping individual pegs. The board used for this example consists of ten slots and nine pegs. The board is numbered and initialized as shown in Figure 3.9. Initially, the nine pegs occupy slots one through nine and slot zero is unoccupied. A peg may jump an adjacent peg (horizontally, or diagonally) into an unoccupied slot. The peg that is jumped is removed from the board. This is similar to capturing a piece by jumping in the game of checkers.

A valid move sequence produced by the program in Code List 3.19 is illustrated in Figure 3.9. The first move in the game is for peg number five to jump over peg number two landing in the empty slot zero. Peg number two is removed from the board and the game continues. The next move is to move peg number seven, jumping over peg number four, and landing in the unoccupied slot two. Peg number four is then removed from the board. The game continues in a similar fashion until there are no more possible moves. At the end of the game in Figure 3.9 three pieces remain on the board: piece number five, piece number six, and piece number eight.

The output of the program is shown in Code List 3.20. The output presents an X if there is a peg remaining at a specific position and a 0 if there is no peg. As seen in the output file at the stage the search is printed out there are three pegs left for each combination. The output is the exhaustive list of all combinations which result in three pegs remaining after six moves. In all cases there are no more additional valid moves. The paths are printed for each solution. Multiple paths give rise to the same final peg distribution for instance

$$[(5,0),(7,2),(0,5),(9,7),(6,8),(1,6)]$$

and

$$[(5,0),(7,2),(9,7),(6,8),(1,6), (0,5)]$$

both result in 00000XX0X0.

One of the problems with the program is the massive amount of data required to store all valid paths which lead to a fixed peg configuration. Consider the problem of expanding the game to the "real" coffee house game which really consists of 14 pegs initially placed on a triangle. If the program is modified to

FIGURE 3.9 A Particular Game Sequence

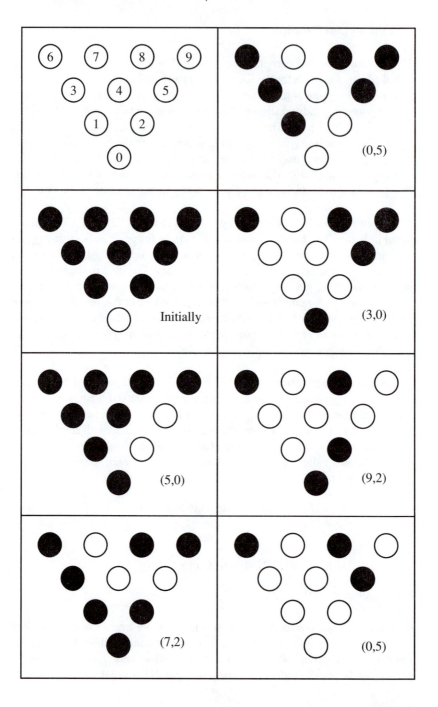

support the new triangle then it requires too much memory to run on most workstations. As a result if the desired problem is to find one path that is optimal a different approach described in the next section must be taken.

3.5.1.1 Bounding a Search Space

In order to minimize the arbitrary expansion of paths for the coffee house game of size 15 the program can be modified to remove any entries in the linked list which duplicate a configuration obtainable via another path. If this approach is taken then only one path will be saved at each point in the iteration for a given intermediate position. This will bound the search space at each iteration and will result in a workable solution. Using a rather unsophisticated argument it is easy to see that the amount of memory is reduced significantly and is realistically bounded. Since each position is represented as a sequence of 15 0's and X's the maximum number of positions under consideration at any time is 2^{15}. For each position only one path is stored instead of the myriad of paths which result in the same position. This approach is used in Problem 3.6 to find a solution for the coffee house game.

Code List 3.19 *Source Code for Game Simulation*

C++ Source Code
#include <iomanip.h>
/* relationship array defining legal movements on triangle */
int rship[][3]={
{0,2,5},
{0,1,3},
{2,5,9},
{2,4,7},
{1,3,6},
{1,4,8},
{3,1,0},
{3,4,5},
{5,2,0},
{5,4,3},
{6,3,1},
{6,7,8},
{7,4,2},
{7,8,9},
{8,7,6},

Code List 3.19 Source Code for Game Simulation (continued)

C++ Source Code

```
            {8,4,1},
            {9,8,7},
            {9,5,2}
};
#define pri
#define printlist
#define max_level 7
#define triangle_size 10
#define testpoint 0
class path
{
        public:
        int src;
        int dest;
        path * next;
        path * prev;
};
class instance
{
        public:
        char config[triangle_size];
        path * pa;
        int elements;
        instance * next;
        instance * prev;
};
class relation
{
        public:
        int del;
        int from;
        relation * next;
```

Code List 3.19 Source Code for Game Simulation (continued)

C++ Source Code

```cpp
};
void path_copy(path * old, path * * pstart)
{
        path *ptemp, *ptemp2;
        *pstart=new path;
        (*pstart)->src=old->src;
        (*pstart)->dest=old->dest;
        (*pstart)->next=NULL;
        (*pstart)->prev=NULL;
        ptemp=old->next;
        ptemp2=*pstart;
        while(ptemp!=NULL)
        {
                ptemp2->next= new path;
                ptemp2->next->prev=ptemp2;
                ptemp2=ptemp2->next;
                ptemp2->src=ptemp->src;
                ptemp2->dest=ptemp->dest;
                ptemp2->next=NULL;
                ptemp=ptemp->next;
        }
}
void path_mem_free(path * list)
{
        path * temp, * temp2;
        if(list==NULL) return;
        temp=list;
        if(temp==NULL) return;
        while(temp->next!=NULL) temp=temp->next;
        while(temp->prev!=NULL)
        {
                temp2=temp;
```

Code List 3.19   Source Code for Game Simulation (continued)

```cpp
C++ Source Code
                temp=temp->prev;
                delete temp2;
        }
        delete temp;
}
void struct_mem_free(instance * list)
{
        instance * temp, * temp2;
        if(list==NULL) return;
        temp=list;
        while(temp!=NULL)
        {
                path_mem_free(temp->pa);
                temp=temp->next;
        }
        temp=list;
        while(temp->next!=NULL) temp=temp->next;
        while(temp->prev!=NULL)
        {
                temp2=temp;
                temp=temp->prev;
                delete temp2;
        }
        delete temp;
}
void add_to_new(relation * rel,instance * old,instance * * new_l, int node)
{
        path * temp_path;
        instance * temp;
        int i;
        if(*new_l == NULL)
        {
```

Code List 3.19 Source Code for Game Simulation (continued)

C++ Source Code

```
                    *new_l = new instance;
                    for(i=0;i<triangle_size;i++)
                              (*new_l)->config[i]=old->config[i];
                    (*new_l)->config[node]=1;
                    (*new_l)->config[rel->del]=0;
                    (*new_l)->config[rel->from]=0;
                    (*new_l)->next=NULL;
                    (*new_l)->prev=NULL;
                    if(old->pa==NULL)
                    {
                              (*new_l)->pa = new path;
                              (*new_l)->pa->src=rel->from;
                              (*new_l)->pa->dest=node;
                              (*new_l)->pa->next=NULL;
                              (*new_l)->pa->prev=NULL;
                    }
                    else
                    {
                              path_copy(old->pa,&temp_path);
                              (*new_l)->pa=temp_path;
                              while(temp_path->next!=NULL)
                                        temp_path=temp_path->next;
                              temp_path->next = new path;
                              temp_path->next->prev=temp_path;
                              temp_path=temp_path->next;
                              temp_path->src=rel->from;
                              temp_path->dest=node;
                              temp_path->next=NULL;
                    }
          }
          else
          {
```

Code List 3.19   Source Code for Game Simulation (continued)

C++ Source Code

```
temp=*new_l;
while(temp->next!=NULL) temp=temp->next;
temp->next=new instance;
for(i=0;i<triangle_size;i++)
        (temp->next)->config[i]=old->config[i];
(temp->next)->config[node]=1;
(temp->next)->config[rel->del]=0;
(temp->next)->config[rel->from]=0;
(temp->next)->next=NULL;
(temp->next)->prev=temp;
temp=temp->next;
if(old->pa==NULL)
{
        temp->pa = new path;
        temp->pa->src=rel->from;
        temp->pa->dest=node;
        temp->pa->next=NULL;
        temp->pa->prev=NULL;
}
else
{
        path_copy(old->pa,&temp_path);
        temp->pa=temp_path;
        while(temp_path->next!=NULL)
                temp_path=temp_path->next;
        temp_path->next = new path;
        temp_path->next->prev=temp_path;
        temp_path=temp_path->next;
        temp_path->src=rel->from;
        temp_path->dest=node;
        temp_path->next=NULL;
}
```

Code List 3.19 Source Code for Game Simulation (continued)

```
C++ Source Code
              }
      }
      void check_move(relation * rel,instance * old,instance * * new_l,int node)
      {
              while(rel!=NULL)
              {
                      if((old->config[rel->from]==1) &&
                              (old->config[rel->del]==1))
                      {
                              add_to_new(rel,old,new_l,node);
                      }
                      rel=rel->next;
              }
      };
      void print_list(instance * list)
      {
              path * pa;
              int i;
      #ifdef printlist
              for(i=0;i<triangle_size;i++)
              {
                      if(list->config[i]==0)
                              cout << "O";
                      else cout << "X";
              };
              cout << " ";
      #endif
              pa = list->pa;
              cout << "[";
              while(pa!=NULL)
              {
                      cout << "(" << pa->src << "," << pa->dest << ")";
```

Code List 3.19 Source Code for Game Simulation (continued)

C++ Source Code

```
                pa=pa->next;
                if (pa) cout << ",";
        }
        cout << "]" << endl;
}
void main()
{
        int i,node,level;
        instance * old_list, * new_list, *tmp_list, *tmp2;
        relation * rel[triangle_size], * temp;
        int RCNT = sizeof(rship)/(3*sizeof(int));
        /* generate data for the initial instance */
        old_list = new instance;
        for(i=0;i<triangle_size;i++) old_list->config[i]=1;
        old_list->config[testpoint]=0;
        old_list->elements=14;
        old_list->next=NULL;
        old_list->prev=NULL;
        old_list->pa=NULL;
        new_list=NULL;
        /*end code for initial instance */
        print_list(old_list);
        /* code to define relationships */
        for(i=0;i<triangle_size;i++) rel[i] = NULL;
        for(i=0; i< RCNT ;i++)
        {
                node=rship[i][0];
                if (rel[node]==NULL)
                {
                        rel[node]=new relation;
                        rel[node]->del=rship[i][1];
```

Code List 3.19 Source Code for Game Simulation (continued)

C++ Source Code

```
                              rel[node]->from=rship[i][2];
                              rel[node]->next=NULL;

                }
                else
                {
                              temp = rel[node];
                              while(temp->next != NULL) temp = temp->next;
                              temp->next = new relation;
                              temp=temp->next;
                              temp->next = NULL;
                              temp->del = rship[i][1];
                              temp->from=rship[i][2];
                }
        };/*end for*/
        /* end code to define relationships */
        for(level=0;level<max_level;level++)
        {
                  tmp2=old_list;
                  while(old_list!=NULL){
                          for(i=0;i<triangle_size;i++)
                          {
                          if(old_list->config[i]==0)
                          /* found candidate for expansion */
                          check_move(rel[i],old_list,&new_list,i);
                          } /* end for */
                          old_list=old_list->next;
                  };
                  /* end do */
                  struct_mem_free(tmp2);
#ifdef pri
                  tmp_list=new_list;
                  while(tmp_list!=NULL)
```

Code List 3.19 Source Code for Game Simulation (continued)

C++ Source Code

```
                {
                        if(level== max_level–2) print_list(tmp_list);
                        tmp_list=tmp_list->next;
                }
#endif
                old_list=new_list;
                new_list=NULL;
        }/* end for(level.... */
}
```

Code List 3.20 Output of Program in Code List 3.19

C++ Output
OXXXXXXXXX []
OOOOOXXOXO [(5,0),(7,2),(0,5),(3,0),(9,2),(0,5)]
XOOOOXOOXO [(5,0),(7,2),(0,5),(3,0),(9,7),(6,8)]
OOOOOXXOXO [(5,0),(7,2),(0,5),(9,2),(3,0),(0,5)]
XOOOOXOOXO [(5,0),(7,2),(0,5),(9,7),(3,0),(6,8)]
XOOOOXOOXO [(5,0),(7,2),(0,5),(9,7),(6,8),(3,0)]
OOOOOXXOXO [(5,0),(7,2),(0,5),(9,7),(6,8),(1,6)]
XOOOOXOOXO [(5,0),(7,2),(9,7),(0,5),(3,0),(6,8)]
XOOOOXOOXO [(5,0),(7,2),(9,7),(0,5),(6,8),(3,0)]
OOOOOXXOXO [(5,0),(7,2),(9,7),(0,5),(6,8),(1,6)]
XOOOOXOOXO [(5,0),(7,2),(9,7),(6,8),(0,5),(3,0)]
OOOOOXXOXO [(5,0),(7,2),(9,7),(6,8),(0,5),(1,6)]
OOOOOXXOXO [(5,0),(7,2),(9,7),(6,8),(1,6),(0,5)]
OXXOOOXOOO [(5,0),(3,5),(9,2),(0,3),(6,1),(8,6)]
OXXOOOOOOX [(5,0),(3,5),(9,2),(0,3),(6,1),(7,9)]
OXXOOOOOOX [(5,0),(3,5),(9,2),(0,3),(7,9),(6,1)]
OXXOOOXOOO [(5,0),(3,5),(9,2),(0,5),(7,9),(9,2)]
OXXOOOOOOX [(5,0),(3,5),(9,2),(7,9),(0,3),(6,1)]
OXXOOOXOOO [(5,0),(3,5),(9,2),(7,9),(0,5),(9,2)]

Code List 3.20 Output of Program in Code List 3.19 (continued)

C++ Output
OXXOOOXOOO [(5,0),(3,5),(0,3),(6,1),(9,2),(8,6)]
OXXOOOOOOX [(5,0),(3,5),(0,3),(6,1),(9,2),(7,9)]
OXXOOOXOOO [(5,0),(3,5),(0,3),(6,1),(8,6),(9,2)]
OXXOOOXOOO [(5,0),(3,5),(0,3),(9,2),(6,1),(8,6)]
OXXOOOOOOX [(5,0),(3,5),(0,3),(9,2),(6,1),(7,9)]
OXXOOOOOOX [(5,0),(3,5),(0,3),(9,2),(7,9),(6,1)]
OOOXOOOXOX [(3,0),(8,1),(0,3),(5,0),(6,1),(0,3)]
XOOXOOOXOO [(3,0),(8,1),(0,3),(5,0),(6,8),(9,7)]
OOOXOOOXOX [(3,0),(8,1),(0,3),(6,1),(5,0),(0,3)]
XOOXOOOXOO [(3,0),(8,1),(0,3),(6,8),(5,0),(9,7)]
XOOXOOOXOO [(3,0),(8,1),(0,3),(6,8),(9,7),(5,0)]
OOOXOOOXOX [(3,0),(8,1),(0,3),(6,8),(9,7),(2,9)]
XOOXOOOXOO [(3,0),(8,1),(6,8),(0,3),(5,0),(9,7)]
XOOXOOOXOO [(3,0),(8,1),(6,8),(0,3),(9,7),(5,0)]
OOOXOOOXOX [(3,0),(8,1),(6,8),(0,3),(9,7),(2,9)]
XOOXOOOXOO [(3,0),(8,1),(6,8),(9,7),(0,3),(5,0)]
OOOXOOOXOX [(3,0),(8,1),(6,8),(9,7),(0,3),(2,9)]
OOOXOOOXOX [(3,0),(8,1),(6,8),(9,7),(2,9),(0,3)]
OXXOOOOOOX [(3,0),(5,3),(6,1),(0,3),(8,6),(6,1)]
OXXOOOXOOO [(3,0),(5,3),(6,1),(0,5),(9,2),(8,6)]
OXXOOOOOOX [(3,0),(5,3),(6,1),(0,5),(9,2),(7,9)]
OXXOOOXOOO [(3,0),(5,3),(6,1),(0,5),(8,6),(9,2)]
OXXOOOOOOX [(3,0),(5,3),(6,1),(8,6),(0,3),(6,1)]
OXXOOOXOOO [(3,0),(5,3),(6,1),(8,6),(0,5),(9,2)]
OXXOOOXOOO [(3,0),(5,3),(0,5),(6,1),(9,2),(8,6)]
OXXOOOOOOX [(3,0),(5,3),(0,5),(6,1),(9,2),(7,9)]
OXXOOOXOOO [(3,0),(5,3),(0,5),(6,1),(8,6),(9,2)]
OXXOOOXOOO [(3,0),(5,3),(0,5),(9,2),(6,1),(8,6)]
OXXOOOOOOX [(3,0),(5,3),(0,5),(9,2),(6,1),(7,9)]
OXXOOOOOOX [(3,0),(5,3),(0,5),(9,2),(7,9),(6,1)]

3.6   Linear Search

A linear search is a search which proceeds in a linear fashion through a list.

The C++ code to perform a linear search on strings is shown in Code List 3.21. The output of the program is shown in Code List 3.22

Code List 3.21 Linear Search Code for Strings

```
C++ Source Code

#include <stdlib.h>
#include <iostream.h>
#include <string.h>


// Initialize the array Note that array must be sorted
char array[][10] = {"Data1", "Data2", "Data3","Data4", "Data5", "Data6",
                    "Data7", "Data8"};


// function used by bsearch to compare data
int compare(const void * i, const void * j)
{
cout << (char *) i << " is compared to ";
cout << (char *) j << endl;
return(strcmp((char *) i,(char *) j));
}


int find(char * key)
{
int * ptr;
size_t number_elements=8;
ptr = (int *) lfind(key,(void *) array,&number_elements,10,compare);
return(ptr!=NULL);
}


void main(void)
{
if(find("Data1")) cout << "Data1 is in list" << endl;
```

Code List 3.21 Linear Search Code for Strings (continued)

C++ Source Code

else cout << "Data1 is not in list" << endl;
if(find("Data12")) cout << "Data12 is in list" << endl;
else cout << "Data12 is not in list" << endl;
 	
}

Code List 3.22 Output of Program in Code List 3.21

C++ Output
Data1 is compared to Data1
Data1 is in list
Data12 is compared to Data1
Data12 is compared to Data2
Data12 is compared to Data3
Data12 is compared to Data4
Data12 is compared to Data5
Data12 is compared to Data6
Data12 is compared to Data7
Data12 is compared to Data8
Data12 is not in list

3.7 Binary Search

The binary search is used in a sorted array to search for an element. The search consists of comparing against the middle of the list and proceeding to search the higher or lower sublist in a recursive fashion.

A binary search is shown in C++ in Code List 3.23. The output is shown in Code List 3.24. A binary search for strings is illustrated in Code List 3.25. The output of the program is shown in Code List 3.25.

Code List 3.23 Binary Search for Integers

C++ Source Code
#include <stdlib.h> #include <iostream.h>

Code List 3.23 Binary Search for Integers (continued)

```cpp
C++ Source Code

// Initialize the array
int array[] = {100,200,50,80,90,600};

// function used by bsearch to compare data
int compare(const void * i, const void * j)
{
return(*(int *) i – *(int *) j);
}

int find(int key)
{
int * ptr;
ptr = (int *) bsearch(&key,array,6,sizeof(int),compare);
return(ptr!=NULL);
}

void main(void)
{
if(find(80)) cout << "80 is in list" << endl;
        else cout << "80 is not in list" << endl;
if(find(81)) cout << "81 is in list" << endl;
        else cout << "81 is not in list" << endl;
}
```

Code List 3.24 Output of Program in Code List 3.23

```
C++ Output
80 is in list
81 is not in list
```

3.8 QuickSort

The quick sort algorithm is a simple yet quick algorithm to sort a list. The algorithm is comprised of a number of stages. At each stage a key is chosen.

Code List 3.25 Binary Search for Strings

```
C++ Source Code
#include <stdlib.h>
#include <iostream.h>
#include <string.h>


// Initialize the array Note that array must be sorted
char array[][10] = {"Data1", "Data2", "Data3","Data4", "Data5", "Data6",
                    "Data7", "Data8"};


// function used by bsearch to compare data
int compare(const void * i, const void * j)
{
cout << (char *) i << " is compared to ";
cout << (char *) j << endl;
return(strcmp((char *) i,(char *) j));
}


int find(char * key)
{
int * ptr;
ptr = (int *) bsearch(key,(void *) array,8,10,compare);
return(ptr!=NULL);
}


void main(void)
{
if(find("Data1")) cout << "Data1 is in list" << endl;
        else cout << "Data1 is not in list" << endl;
if(find("Data12")) cout << "Data12 is in list" << endl;
        else cout << "Data12 is not in list" << endl;
}
```

The algorithm starts at the left of the list until an element is found which is greater than the key. Starting from the right, an element is searched for which

---

Code List 3.26 Output of Program in Code List 3.25

C++ Output
Data1 is compared to Data5
Data1 is compared to Data3
Data1 is compared to Data2
Data1 is compared to Data1
Data1 is in list
Data12 is compared to Data5
Data12 is compared to Data7
Data12 is compared to Data6
Data12 is not in list

is less than the key. When both the elements are found they are exchanged. After a number of iterations the list will be divided into two lists. One list will have all its elements less than or equal to the key and the other list will have all its elements greater than or equal to the key. The two lists created are then each sorted by the same algorithm.

The internal details of a quicksort algorithm are shown in the C++ program in Code List 3.27. The output of the program is shown in Code List 3.28.

A number of different approaches can be used to determine the key. The quicksort algorithm in this section uses the median of three approach. In this approach a key is chosen for each search segment.

The key is given as the median of three on the bounds of the segment. For instance, in Code List 3.28, the initial segment to sort contains 18 elements, indexed 0–17. The first key is determined by the calculation

$$key = \left\lfloor \frac{(x[0] + x[8] + x[17])}{3} \right\rfloor$$

$$= \left\lfloor \frac{(300 + 455 + 12)}{3} \right\rfloor = \left\lfloor \frac{767}{3} \right\rfloor = 255 \qquad \text{(3.1)}$$

After the comparisons two lists are formed. In this case the lists are 0–8 and 9–17. Every element in the first list will be less than or equal to the key 255 and everything in the second list will be greater than or equal to 255. The two new lists can be sorted in parallel. This example is sequential code so that the second list 9–17 is dealt with first.

The comparisons occurring within the first list is illustrated in Code List 3.29. Two comparisons can be done in parallel. Starting from the left a search is made for the first element greater than 255. In this case the first element satisfies that criteria.

Starting from the right a search is made for the first element that is less than 255. In this case it is the last element. At this point the two elements are exchanged in the list which results in the second list in Code List 3.29. Continuing in this manner proceeding from the left the next element in the list is searched for which is greater than 255. In this case it is the third element in the list, 415. Proceeding from the right the first element less than 255 found is 100. Again, 100 and 415 are exchanged resulting in the third list. Eventually the two left and right pointers overlap indicating that the list has been successfully sorted about the key.

C++ also provides a quicksort operator which performs the median of three sort. This is illustrated for strings is illustrated in Code List 3.34. The output of the program is shown in Code List 3.35 A quicksort C++ program for doubles is shown in Code List 3.30 The output is shown in Code List 3.31. A quicksort program for integers is shown in Code List 3.32. The output is shown in Code List 3.33.

Code List 3.27 QuickSort C++ Program

```
C++ Source Code

#include <iostream.h>

// Data for the sort algorithm
int data[] = {300,200,415,406,433,89,42,767,
                    455,321,309,1045,114,87,-6,89,100,12};

// This is the class for the subsets of the data to be sorted
class subset
        {
        public:
        int left;
        int right;
        subset * next;
        };
```

Code List 3.27 QuickSort C++ Program (continued)

C++ Source Code

```cpp
// The primary class
class array
        {
        subset * list;
        public:
        void print_data();
        int get_key(subset * list);
        void print_key(subset * list);
        void exchange(int i, int j);
        void compare(int *i, int *j, int k);
        void quick_sort();
        array() { list = new subset;
                list->left=0;
                list->right=sizeof(data)/sizeof(int)–1;
                list->next=NULL;
                }
        };


// This functions prints the value of the data
void array::print_data()
        {
        int i;
        for(i=0;i<sizeof(data)/sizeof(int);i++) cout << data[i] << " ";
        cout << endl;
        }



// This returns the key for the first bounds in the list
int array::get_key(subset * list)
        {
        return (data[list->left]+data[list->right]+
                data[(list->left+list->right)/2])/3;
```

Code List 3.27 QuickSort C++ Program (continued)

C++ Source Code

```
        }

// This prints the value of the key for the pointer passed
void array::print_key(subset * list)
        {
        cout << "Present key = " << get_key(list) << endl;
        }

void array::exchange(int i, int j)
        {
        int tmp;
        tmp = data[i];
        data[i]=data[j];
        data[j]=tmp;
        }

// This routine compares data within the bounds to the key k
// This routine performs at most one exchange
void array::compare(int *i,int *j, int k)
        {
        int m=*i,n=*j;
        for(;m<*j;m++) if(data[m]>k) break;
        for(;n>*i;n--) if(data[n]<k) break;
        if(m<n) exchange(m,n);
        *i=m;
        *j=n;
        }




void array::quick_sort()
        {
```

Code List 3.27 QuickSort C++ Program (continued)

C++ Source Code

```cpp
while(list!=NULL)
{
int i,j,k;
subset * tmp=list->next;
subset * newl;
i=list->left;
j=list->right;
k=get_key(list);
cout << endl << "Working on list " << i << " " << j << endl;
print_key(list);
print_data();
while(i<j) compare(&i,&j,k);


if(list->left < j)
        {
        cout << "create new list " << list->left
                << " " << j << endl;
        newl = new subset;
        newl->left = list->left;
        newl->right = j;
        newl->next=tmp;
        tmp=newl;

        }
if(list->right > i)
        {
        cout << "create new list " << i
                << " " << list->right << endl;
        newl = new subset;
        newl->left= i;
        newl->right = list->right;
```

Code List 3.27 QuickSort C++ Program (continued)

```
C++ Source Code

                    newl->next=tmp;
                    tmp=newl;
                    }

            print_data();
            delete list;
            list=tmp;
            }
            }
void main()
            {
            array x;
            x.quick_sort();
            }
```

Code List 3.28 Output of Program in Code List 3.27

```
C++ Output

Working on list 0 17
Present key = 255
300 200 415 406 433 89 42 767 455 321 309 1045 114 87 –6 89 100 12
create new list 0 8
create new list 9 17
12 200 100 89 –6 89 42 87 114 321 309 1045 455 767 433 406 415 300

Working on list 9 17
Present key = 462
12 200 100 89 –6 89 42 87 114 321 309 1045 455 767 433 406 415 300
create new list 9 15
create new list 16 17
12 200 100 89 –6 89 42 87 114 321 309 300 455 415 433 406 767 1045
```

Code List 3.28 Output of Program in Code List 3.27 (continued)

C++ Output

Working on list 16 17

Present key = 859

12 200 100 89 –6 89 42 87 114 321 309 300 455 415 433 406 767 1045

12 200 100 89 –6 89 42 87 114 321 309 300 455 415 433 406 767 1045

Working on list 9 15

Present key = 394

12 200 100 89 –6 89 42 87 114 321 309 300 455 415 433 406 767 1045

create new list 9 11

create new list 12 15

12 200 100 89 –6 89 42 87 114 321 309 300 455 415 433 406 767 1045

Working on list 12 15

Present key = 425

12 200 100 89 –6 89 42 87 114 321 309 300 455 415 433 406 767 1045

create new list 12 13

create new list 14 15

12 200 100 89 –6 89 42 87 114 321 309 300 406 415 433 455 767 1045

Working on list 14 15

Present key = 440

12 200 100 89 –6 89 42 87 114 321 309 300 406 415 433 455 767 1045

12 200 100 89 –6 89 42 87 114 321 309 300 406 415 433 455 767 1045

Working on list 12 13

Present key = 409

12 200 100 89 –6 89 42 87 114 321 309 300 406 415 433 455 767 1045

12 200 100 89 –6 89 42 87 114 321 309 300 406 415 433 455 767 1045

Working on list 9 11

Present key = 310

Code List 3.28 Output of Program in Code List 3.27 (continued)

C++ Output
12 200 100 89 –6 89 42 87 114 321 309 300 406 415 433 455 767 1045
create new list 9 10
12 200 100 89 –6 89 42 87 114 300 309 321 406 415 433 455 767 1045
Working on list 9 10
Present key = 303
12 200 100 89 –6 89 42 87 114 300 309 321 406 415 433 455 767 1045
12 200 100 89 –6 89 42 87 114 300 309 321 406 415 433 455 767 1045
Working on list 0 8
Present key = 40
12 200 100 89 –6 89 42 87 114 300 309 321 406 415 433 455 767 1045
create new list 0 1
create new list 2 8
12 –6 100 89 200 89 42 87 114 300 309 321 406 415 433 455 767 1045
Working on list 2 8
Present key = 101
12 –6 100 89 200 89 42 87 114 300 309 321 406 415 433 455 767 1045
create new list 2 6
create new list 7 8
12 –6 100 89 87 89 42 200 114 300 309 321 406 415 433 455 767 1045
Working on list 7 8
Present key = 171
12 –6 100 89 87 89 42 200 114 300 309 321 406 415 433 455 767 1045
12 –6 100 89 87 89 42 114 200 300 309 321 406 415 433 455 767 1045
Working on list 2 6
Present key = 76
12 –6 100 89 87 89 42 114 200 300 309 321 406 415 433 455 767 1045
create new list 3 6

Code List 3.28 Output of Program in Code List 3.27 (continued)

C++ Output
12 –6 42 89 87 89 100 114 200 300 309 321 406 415 433 455 767 1045
Working on list 3 6
Present key = 92
12 –6 42 89 87 89 100 114 200 300 309 321 406 415 433 455 767 1045
create new list 3 5
12 –6 42 89 87 89 100 114 200 300 309 321 406 415 433 455 767 1045
Working on list 3 5
Present key = 88
12 –6 42 89 87 89 100 114 200 300 309 321 406 415 433 455 767 1045
create new list 4 5
12 –6 42 87 89 89 100 114 200 300 309 321 406 415 433 455 767 1045
Working on list 4 5
Present key = 89
12 –6 42 87 89 89 100 114 200 300 309 321 406 415 433 455 767 1045
12 –6 42 87 89 89 100 114 200 300 309 321 406 415 433 455 767 1045
Working on list 0 1
Present key = 6
12 –6 42 87 89 89 100 114 200 300 309 321 406 415 433 455 767 1045
–6 12 42 87 89 89 100 114 200 300 309 321 406 415 433 455 767 1045

Code List 3.29 QuickSort Comparison

Comparisons on First List 0–17
Working on list 0 17
Present key = 255
300 200 415 406 433 89 42 767 455 321 309 1045 114 87 –6 89 100 12
12 200 415 406 433 89 42 767 455 321 309 1045 114 87 –6 89 100 300
12 200 100 406 433 89 42 767 455 321 309 1045 114 87 –6 89 415 300

Code List 3.29 QuickSort Comparison (continued)

Comparisons on First List 0–17
12 200 100 89 433 89 42 767 455 321 309 1045 114 87 –6 406 415 300
12 200 100 89 –6 89 42 767 455 321 309 1045 114 87 433 406 415 300
12 200 100 89 –6 89 42 87 455 321 309 1045 114 767 433 406 415 300
12 200 100 89 –6 89 42 87 114 321 309 1045 455 767 433 406 415 300

Code List 3.30 QuickSort For Double Types

C++ Source Code

```cpp
#include <stdio.h>
#include <stdlib.h>
#include <string.h>
#include <iostream.h>
int user_sort(const void *a, const void *b);
double age[]={45.0,25.5,12,29,37,37,41.1};
void main()
        {
        int i;
        qsort((void *)age,7,sizeof(double),user_sort);
        for(i=0;i<7;i++) cout << age[i] <<endl;
        }
int user_sort(const void *a, const void *b)
        {
        if( *(double *) a < *(double *)b) return –1;
        if( *(double *) a > *(double *)b) return 1;
return 0;
        }
```

Code List 3.31 Output for Program in Code List 3.30

C++ Output
12
25.5

Code List 3.31 Output for Program in Code List 3.30 (continued)

C++ Output
29
37
37
41.1
45

Code List 3.32 QuickSort Program for Integers

```
C++ Source Code

#include <stdio.h>
#include <stdlib.h>
#include <string.h>
#include <iostream.h>
int user_sort(const void *a, const void *b);
int age[]={4,14,7,34,23,26,43};
void main()
        {
        int i;
        qsort((void *)age,7,sizeof(int),user_sort);
        for(i=0;i<7;i++) cout << age[i] <<endl;
        }
int user_sort(const void *a, const void *b)
        {
        return( *(int *)a - *(int *)b);
        }
```

Code List 3.33 Output for Program in Code List 3.32

C++ Output
4
7
14

Code List 3.33 Output for Program in Code List 3.32 (continued)

C++ Output
23
26
34
43

Code List 3.34 QuickSort Program

```cpp
#include <stdio.h>
#include <stdlib.h>
#include <string.h>
#include <iostream.h>
int user_sort(const void *a, const void *b);
char names[][10]={"Jones","Gaede","Wells","Nichols"};
void main()
        {
        int i;
        qsort((void *)names,4,10,user_sort);
        for(i=0;i<4;i++) cout << names[i] <<endl;
        }
int user_sort(const void *a, const void *b)
        {
        return(strcmp((char *)a,(char *)b));
        }
```

Code List 3.35 Output of Program in Code List 3.34

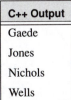

C++ Output
Gaede
Jones
Nichols
Wells

3.9 Binary Trees

A binary tree is a common data structure used in algorithms. A typical *class* supporting a binary tree is

class tree

{

public:

int key;

tree * left;

tree * right;

}

A binary tree is *balanced* if for every node in the tree the height of the left and right subtrees are within one.

3.9.1 Traversing the Tree

There are a number of algorithms for traversing a binary tree given a pointer to the root of the tree. The most common strategies are *preorder*, *inorder*, and *postorder*. The *preorder* strategy visits the root prior to visiting the left and right subtrees. The *inorder* strategy visits the left subtree, the root, and the right subtree. The *postorder* strategy visits the left subtree, the right subtree, followed by the root. These strategies are recursively invoked.

3.10 Hashing

Hashing is a technique in searching which is commonly used by a compiler to keep track of variable names; however, there are many other useful applications which use this approach. The idea is to use a hash function, $h(E)$, on elements, E, to assist in locating an element. For instance a dictionary might be defined using an array of twenty six pointers, $D[26]$. Each pointer points to a linked list of data for the specific letter of the alphabet. The hashing function on the string simply returns the number of the letter of the alphabet minus one of the first characters in the string:

$$h(ace) = 0 \qquad h(zebra) = 25 \qquad \text{(3.2)}$$

There are two major operations which need to be supported for the hash table created:

- search for an element
- search for an element and insert the element if not found
- indicate if the hash table is full

The idea of hashing is to simplify the search process so the hashing function should be simple to calculate. Additionally, there should be a simple way to locate the data, referred to as resolving *collisions*, once the hash function is evaluated.

3.11   Simulated Annealing

The simulated annealing algorithm is illustrated in Figure 3.10. The goal of simulated annealing is to attempt to find an optimum to a large-scale problem which typically cannot be found by conventional means. The solution is sought by iterating and evaluating a cost at each stage. The algorithm maintains a concept of a temperature. When the temperature is high the algorithm will be likely to accept a higher cost solution. When the temperature is very low the algorithm will almost always only accept solutions of lower cost. The temperature begins high and is cooled until an equilibrium is reached. By allowing the initial temperature to be high the algorithm will be allowed to "climb hills" to seek a global optimum. Without this feature it is possible to be trapped in a local minimum. This is illustrated in Figure 3.12. By allowing the function to move to a higher value it is able to climb over the hill and find the global minimum.

Simulated annealing is applied to the square packing problem described in the next section. This illustrates the difficulty and complexity of searching in general problems.

3.11.1   The Square Packing Problem

The square packing problem is as follows:

Given a list of squares (integer sides) determine the smallest square which includes the list of squares in a nonoverlapping manner.

A given instance for the square packing problem is shown in Figure 3.11. For this figure the list of squares provided have sides

$$1,1,1,1,1,2,3,3,3,3,6$$

An optimal solution as shown in the figure packs the squares into a 9x9 square. A C++ source program implementing the simulated annealing algorithm for the square packing problem is shown in Code List 3.36. The output of the program is shown in Code List 3.37.

3.11.1.1 Program Description

This section describes the program. The description begins with the start of the file and proceeds forward.

The program includes a number of files to support the functions in the program. Of importance here is the inclusion of <sys/time.h>. This is machine dependent. This program may have to be modified to run on different platforms. At issue is the conformance to *drand48()* and associated functions as well as the *time* structure format.

The function *drand48()* returns a double random number satisfying

$$0 \le drand48 < 1 \tag{3.3}$$

srand48() is used to seed the random number generator. The defined constants are shown in Table 3.1.

TABLE 3.1 Program Constants

Constant	Meaning
NO_SQUARES	The number of squares in the problem
SQUARE_SIZE_LIMIT	The maximum size of the square. The squares that are generated will have sides from 1 to SQUARE_SIZE_LIMIT. This is used when the initial linked list is generated with random square sides.
INITIAL_TEMPERATURE	The initial temperature in the simulated annealing process.
R	The temperature cooling ratio. The temperature is cooled by this factor each time NO_STEPS have been performed.
NO_ITERATIONS	The number of times to cool. This is the number of times the temperature is reduced by a factor of R.
NO_STEPS	This is the number of steps in the algorithm to perform at the fixed temperature.

TABLE 3.1 Program Constants (continued)

Constant	Meaning
PLUS	This is the representation for the PLUS operator which is used to represent when blocks are placed on top of each other.
TIMES	This is the representation for the TIMES operator which is used to represent when blocks are placed next to each other.
TEST	When this is defined the test data, for which the optimal solution is known, is used.

The representation used in the program for placing the squares is a stacked base approach. Squares placed on top of each other are noted with a +. Squares placed next to each other are noted with a *.

The notation 1 2 * means square 2 to the right of 1. The notation 1 2 + means square 1 on top of 2. The notation is unraveled in a stack base manner so to evaluate the meaning of 0 1 2 3 *4 + * + you push each of the elements on the stack and when you encounter an operation you remove two elements from the stack and replace it with the modified element. The array results in the operation in Table 3.2:

TABLE 3.2 Interpreting Representation

Representation	Meaning
0 1 2 3 * 4 + * +	Original Array
0 1 5 4 + * +	Block 5 created which is composed of block 2 next to 3
0 1 6 * +	Block 6 created which is composed of block 5 on top of 4
0 7 +	Block 7 created which is block 1 next to 6
8	Block 8 created which is block 0 on top of 7

A possible notation, for instance, for Figure 3.11, is

0	1	2	+	*	5	+	6	+	8	9	*	10	*	+	3	4	*	7	+	*

This would represent the square packed into the 9x9 square. Notice that each of the blocks above contain a number or an operation. The program elects to define the + operation as the number NO_SQUARES and the TIMES operation as the NO_SQUARES+1. As a result the valid representations will be the numbers 0–12.

Two stacks are defined in the program, one to store the current *x* width of a box and the current *y* width. This is needed because when you combine squares of different sizes you end up with a rectangle. If you combine a 1x1 with a 2x2 you will end up with a 3x2 or a 2x3.

The test data is initially stored as

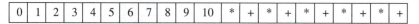

The program starts with the array and perturbs it by replacing it with a neighboring array and evaluating the cost of the string. The *calculate_cost()* function calculates the cost of a given array.

To calculate a neighboring array the algorithm selects a random strategy. This is a required aspect to simulated annealing. The neighboring strategy must be random. The strategy is described in Table 3.3.

TABLE 3.3 Neighbor Solution Strategy

Operation	Description
A_op_to_op_A()	Swap an operation with an element. For instance replace 10 + with + 10.
op_A_to_A_op()	Swap an operation with an element. For instance replace + 10 with 10 +.
AB_to_BA	Exchange two elements. For instance replace 4 5 + to 5 4 +.
switch_op()	switch two operators in the sequence. For instance replace 4 5 * + with 4 5 + *.
ABC_op_to_AB_op_C()	replace a sequence of three elements followed by an operation to two elements followed by the operation followed by the last element. For instance replace 2 4 3 5 + 6 with 2 4 3 + 5 6.
	Notice this is similar to A_op_to_op_A().

There are certain representations which are not valid that are handled by the program. For instance

3 4 * 5 +

cannot be replaced with

3 * 4 5 +

because you need two elements for each operation you run into. In general at any point in the array the number of elements to that point must exceed the number of operations to that point by 1. The program ensures that only valid perturbations are considered.

The output of the program is shown in Code List 3.37. The program found an optimal solution. Since the program is a random program it may not find the optimal solution each time. The program also doesn't output the square number but rather the size of the size. This increases the readability of the solution. The solution to the problem is not unique.

Code List 3.36 Simulated Annealing

C++ Source Code
| ```
#include <math.h>
#include <stdlib.h>
#include <sys/time.h>
#include <iostream.h>
extern double drand48();
extern long lrand48(/*long*/);
extern int rand();
extern void srand48(long seedval);
#define NO_SQUARES 11
#define SQUARE_SIZE_LIMIT 6
#define INITIAL_TEMPERATURE 70.0
#define R 0.8
#define NO_ITERATIONS 20
#define NO_STEPS 10000
#define PLUS NO_SQUARES
#define TIMES NO_SQUARES+1
#define TRUE 1
#define FALSE 0
#define TEST
int op[2*NO_SQUARES–1];
``` |
```

FIGURE 3.10 Generic Simulated Annealing Algorithm

begin

    Start with Initial Solution, $S = S_0$

    Start with Initial Temperature, $T = T_0$

    while(not satisfied) **do**

        begin

            while (not in equilibrium) **do**

                begin

                    S' is some random neighbor of S

                    Calculate Cost differential:

$$(\delta = Cost(S') - Cost(S))$$

                    Assign a probability, $prob$,

$$prob = min\{1, e^{-\delta/t}\}$$

                    if $random(0, 1) \leq prob$ **then** $S = S'$

                end;

            Update Temperature T;

        end;

    Output solution S

Code List 3.36 Simulated Annealing (continued)

C++ Source Code
int present_op[2*NO_SQUARES–1];
int data[NO_SQUARES];

FIGURE 3.11 A Given Instance of the Square Packing Problem

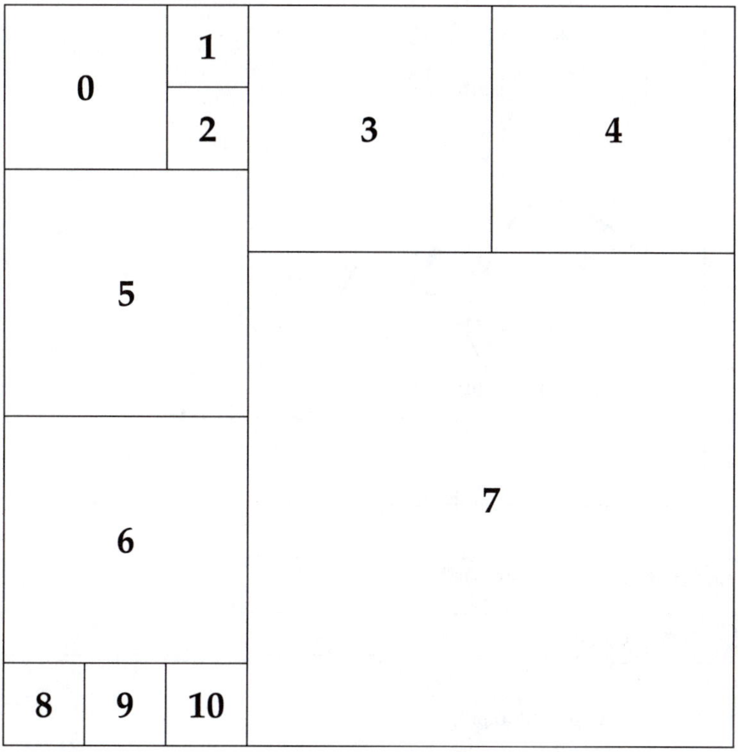

Code List 3.36 Simulated Annealing (continued)

C++ Source Code
int stack_x[NO_SQUARES], stack_y[NO_SQUARES];
int stack_pointer;
int x0_val, y0_val;
int x1_val, y1_val;
#ifdef TEST
int test_set[NO_SQUARES]={1,1,1,1,1,2,3,3,3,3,6};

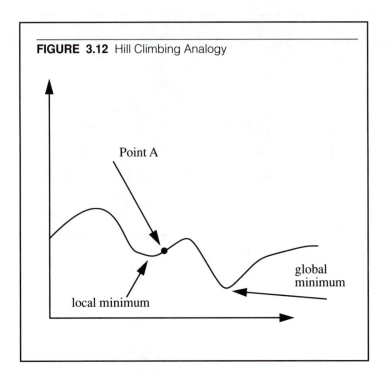

FIGURE 3.12 Hill Climbing Analogy

Code List 3.36 Simulated Annealing (continued)

C++ Source Code

```cpp
#endif
// list structure passed to progam
class square_list {
        public:
        int side;
        square_list * next;
};
void get_list_start(square_list * * list)
{
int i;
struct square_list * build_list, * list_start;
build_list=new square_list;
list_start = build_list;
for(i=0;i<NO_SQUARES;i++)
```

Code List 3.36 Simulated Annealing (continued)

C++ Source Code

```cpp
            {
#ifndef TEST
            build_list->side = (lrand48()>>4)%(SQUARE_SIZE_LIMIT)+1;
#endif
#ifdef TEST
            build_list->side = test_set[i];
#endif
            build_list->next = new square_list;
            build_list=build_list->next;
            }
#ifndef TEST
            build_list->side = rand()%(SQUARE_SIZE_LIMIT)+1;
#endif
            build_list->next = NULL;
            *list = list_start;
}


/***********************************************************/
/* create initial operation array */
/* operations are alternately chosen PLUS and TIMES */
void create_operation_array()
{
            int i,j;
            for(i=0;i<NO_SQUARES;i++) op[i]=i;
            for(j=0;j<NO_SQUARES;j++) op[i++]=PLUS + i%2;
}


/***********************************************************/
/* remove data from stack */
void pop()
{
            stack_pointer--;
```

Code List 3.36 Simulated Annealing (continued)

```cpp
            x1_val=x0_val;
            y1_val=y0_val;
            x0_val=stack_x[stack_pointer];
            y0_val=stack_y[stack_pointer];
}
/*********************************************************/
/* push data on to stack */
void push(int x, int y)
{
            stack_x[stack_pointer]=x;
            stack_y[stack_pointer]=y;
            stack_pointer++;
}
/*********************************************************/
/* pieces side by side */
void merge_times()
{
            if(y0_val <= y1_val) y0_val = y1_val;
            x0_val += x1_val;
}
/*********************************************************/
/* pieces on top of each other */
void merge_plus()
{
            if(x0_val <= x1_val) x0_val = x1_val;
            y0_val += y1_val;
}
/*********************************************************/
/* calculate the cost of op array */
int calculate_cost()
{
            int i;
```

Code List 3.36 Simulated Annealing (continued)

C++ Source Code

```
            stack_pointer=0;
            for(i=0;i<2*NO_SQUARES-1;i++)
            {
                    switch(op[i])
                    {
                    case PLUS:
                            pop();
                            pop();
                            merge_plus();
                            push(x0_val,y0_val);
                            break;

                    case TIMES:
                            pop();
                            pop();
                            merge_times();
                            push(x0_val,y0_val);
                            break;

                            /* data */
                    default:
                            push(data[op[i]],data[op[i]]);
                            break;
                    }
            }
            pop();
            if(x0_val>=y0_val) return(x0_val*x0_val);
            else return(y0_val*y0_val);
}
//***************************************************/
// function to determine if an item is data or an operation */
int is_data(int x)
```

Code List 3.36 Simulated Annealing (continued)

C++ Source Code

```cpp
{
        if((op[x]!=TIMES)&&(op[x]!=PLUS)) return(TRUE);
        else return(FALSE);
}
//****************************************************/
// neighbor solution calculations */
void switch_op()
{
        int i,j,k,loc;
        k = -1;


        /* choose random operator */
        j = lrand48()%(NO_SQUARES-1);

        /* search for location */
        for(i=0;i<2*NO_SQUARES-1;i++)
        {
                if((op[i]==PLUS)||(op[i]==TIMES))
                {
                        k++;
                        if(k==j) loc=i;

                }
        }

        // swap */
        if(op[loc]==PLUS) op[loc]=TIMES;
        else op[loc]=PLUS;
}
/****************************************************/
/* neighbor solution calculations */
void ABC_op_to_AB_op_C()
```

Code List 3.36 Simulated Annealing (continued)

C++ Source Code

```
{
        int i,j,k,temp;
        k=0;
        for(i=0;i<2*NO_SQUARES–4;i++)
                if((is_data(i))&&(is_data(i+1))&&(is_data(i+2))
                        &&(!is_data(i+3))) k++;
        if(k==0) return;
        j=lrand48()%k;
        k=0;
        for(i=0;i<2*NO_SQUARES–4;i++)
                if((is_data(i))&&(is_data(i+1))&&(is_data(i+2))
                        &&(!is_data(i+3))&&(k++==j))
        {
                temp=op[i+2];
                op[i+2]=op[i+3];
                op[i+3]=temp;
                return;
        }
}
/**********************************************************/
/* neighbor solution calculations */
void op_A_to_A_op()
{
        int i,j,k,temp;
        k=0;
        for(i=0;i<2*NO_SQUARES–2;i++)
                if((!is_data(i))&&(is_data(i+1))) k++;
        if(k==0) return;
        j=lrand48()%k;
        k=0;
        for(i=0;i<2*NO_SQUARES–2;i++)
                if((!is_data(i))&&(is_data(i+1))&&(k++==j))
```

Code List 3.36 Simulated Annealing (continued)

C++ Source Code

```
        {
                temp=op[i];
                op[i]=op[i+1];
                op[i+1]=temp;
                return;
        }
}
/*****************************************************/
/* neighbor solution calculations */
void A_op_to_op_A()
{
        int i,j,k,temp;
        int operations=0;
        int data_item=0;
        k=0;
        for(i=0;i<2*NO_SQUARES-3;i++)
        {
        if(is_data(i)) data_item++; else operations++;
        if((is_data(i))&&(!is_data(i+1))
                &&(data_item>=(operations+3))) k++;
        }
        if(k==0) return;

        j=lrand48()%k;
        k=0;
        data_item=0;
        operations=0;
        for(i=0;i<2*NO_SQUARES-3;i++) {
        if(is_data(i)) data_item++; else operations++;
        if((is_data(i))&&(!is_data(i+1))
                &&(data_item>=(operations+3))&&(k++==j))
        {
```

Code List 3.36 Simulated Annealing (continued)

C++ Source Code

```cpp
                temp=op[i];
                op[i]=op[i+1];
                op[i+1]=temp;
                return;
        }
        }
}
/********************************************************/
/* neighbor solution calculations */
void AB_to_BA()
{
        int i,j,k,m,n;
        int k1,temp;
        i=lrand48()%NO_SQUARES;
        j=lrand48()%NO_SQUARES;
        while(i==j) i=lrand48()%NO_SQUARES;
        k1=0;
        for(k=0;k<2*NO_SQUARES-1;k++)
        {
                if(is_data(k))
                {
                        if(i==k1) m=k;
                        if(j==k1) n=k;
                        k1++;
                }
        }
        temp=op[m];
        op[m]=op[n];
        op[n]=temp;
}
/********************************************************/
/* sample data for which optimal is known = 81 */
```

Code List 3.36 Simulated Annealing (continued)

C++ Source Code

```
void sample_data()
{
        int i;
        for(i=0;i<NO_SQUARES;i++) data[i]=2;
}
/****************************************************/
/* randomly select neighbor of op */
void neighbor_solution()
{
        switch((lrand48()>>4)%5)
        {
        case 4: A_op_to_op_A();
                break;
        case 3:
                op_A_to_A_op();
                break;
        case 2:
                AB_to_BA();
                break;
        case 1:
                switch_op();
                break;
        case 0:
                ABC_op_to_AB_op_C();
                break;
        default:
                break;
        }
}


/****************************************************/
/* function to accept neighbor */
```

Code List 3.36 Simulated Annealing (continued)

C++ Source Code

```cpp
void accept_neighbor()
{
        int i;
        for(i=0;i<2*NO_SQUARES-1;i++) present_op[i]=op[i];
}


/*****************************************************/
/* print final output */
void print_results(int optimal_cost)
        {
        int i;
        cout << "Calculated cost " << optimal_cost << endl;
        for(i=0;i<2*NO_SQUARES-1;i++)
                { switch(present_op[i])
                        {
                        case PLUS: cout << "+ ";
                         break;
                        case TIMES: cout << "* ";
                         break;
                        default:
                        cout << data[present_op[i]] << " ";
                         break;
                        }
                }
        cout << endl;
        }


/*****************************************************/
/* main program */
main()
{
        int i,j,k;
```

Code List 3.36 Simulated Annealing (continued)

C++ Source Code

```cpp
class square_list * list_start;
int optimal_cost, random_neighbor_cost;
int cost;
double temperature=INITIAL_TEMPERATURE;
double probability,random_0_1;
/* set random number generator */
struct timeval tp;
struct timezone tzp;
gettimeofday(&tp,&tzp);
srand48(tp.tv_usec);
/* get external list */
get_list_start(&list_start);
/* create operation array for neighbor solution */
create_operation_array();
/* create present active optimal operation_array */
for(i=0;i<2*NO_SQUARES-1;i++) present_op[i]=op[i];
/* get data */
i=0;
while(list_start!=NULL)
{
        data[i++]=list_start->side;
        list_start=list_start->next;
}
/* use sample data from handout if defined at
        beginning of program*/
optimal_cost=calculate_cost();
/* perform annealing */
for(j=0;j<NO_ITERATIONS;j++)
{
        for(i=0;i<NO_STEPS;i++)
        {
```

Code List 3.36 Simulated Annealing (continued)

C++ Source Code

```
                            for(k=0;k<2*NO_SQUARES−1;k++)
                            op[k]=present_op[k];
                            neighbor_solution();
                            cost=calculate_cost();
                            if(cost<=optimal_cost)
                            {
                            accept_neighbor();
                            optimal_cost=cost;
                            }
                            else {
                            probability =
                            exp(−(cost-optimal_cost)/temperature);
                            random_0_1 = drand48();
                            if(random_0_1 <= probability)
                            {
                            accept_neighbor();
                            optimal_cost=cost;
                            };
                            }
                      }
                 temperature*=R;
           }
        print_results(optimal_cost);
        return 1;
}
```

Code List 3.37 Output of Program in Code List 3.36

Output of Program in Code List 3.36
Calculated cost 81
1 1 + 2 * 1 1 * 1 * 3 + 3 + + 3 3 * 6 + *

3.12 Problems

(**3.1**) [Pointers, Dynamic Memory Allocation] Write a C++ program to invert a 3x3 matrix with floating point elements. Your program should only declare triple pointers in *main( )*. Every declaration in *main( )* must be of the form: type * * * variable. This also applies to any loop variables needed. No other variables outside of *main()* should be declared (you can use classes outside of *main())*. Any memory allocated with *new* should be removed with *delete*. Input the matrix using the *cin* operator and output the results using the *cout* operator. If the matrix is not invertible you should print "Matrix not Invertible".

(**3.2**) [Dynamic Memory Allocation, FIFO] Write a C++ program to implement a FIFO stack which allocates space dynamically. The size of the stack should increase dynamically (via *new*) with each push operation and decrease (via *delete*) with each pop operation. Support an operation to print the data presently on the stack.

(**3.3**) [Linked Lists] Write a C++ program to maintain a linked lists of strings. The program should support an operation to count the number of entries in the linked list which match a specific string.

(**3.4**) [Linked Lists, Sorting] Write an operation for the program in Problem 3.3 which will sort the linked list in alphabetical order.

(**3.5**) [Linked Lists] Write a general linked list C++ program which supports operations to

- Combine two lists
- Copy a list.
- Split a list at a specific location into two lists

  Make sure you handle all the special cases associated with the start and end of a list.

(**3.6**) [Bounding] Modify the coffee house game program to find a solution where the triangle dimension is 15. The program should use a bounding technique which results in unique intermediate peg locations at each iteration.

(**3.7**) [Merging Sorted Linked Lists] Write a C++ program to merge two separate sorted lists into one sorted list. Calculate the order of your algorithm in terms of the size of the input list, n.

(**3.8**) [Binary Trees] Write a C++ program which is passed a pointer to a binary tree and prints out the keys traversed via *preorder*, *postorder* and *inorder* strategies. Assume your tree class is defined as

```
class tree
{
public:
int key;
tree * left;
tree * right;
}
```

(3.9) [Balanced Trees] Write a C++ program which inserts an element any-where into a balanced tree and results in a tree structure which is still bal-anced. Assume your tree class is the one defined in Problem 3.8.

(3.10) [Balanced Trees] Write a C++ program which deletes an element with a specific key from a balanced tree and results in a tree structure which is still balanced. Assume your tree class is the one defined in Problem 3.8.

(3.11) [Balanced Trees] Write a C++ program which maintains a sorted key list in a balanced binary tree. You should support insertion and deletion of elements in the tree. For this problem the definition of sorted means that at each node in the tree every element in the left subtree is less than or equal to the root key of the subtree and every element in the right subtree is greater than or equal to the root key of the subtree. After insertions and deletions the tree should be balanced. Assume your tree class is the one defined in Problem 3.8.

(3.12) [Order] Calculate the number of operations in terms of the size of the tree for the performance of the algorithm in Problem 3.10.

(3.13) [Hashing — Difficult] Consider a linked list structure which supports the concept of an element with a number of friends:

```
class element
{
public:
char data[100];
element * f1;
element * f2;
element * f3;
}
```

Consider a number of strings, say 2000, to be placed in classes of this nature. Develop a hashing algorithm which will use the fact that an element has three friends to determine the location of the string given only a pointer to a root element. Support the hashing functions to search and insert strings into the table. Try to characterize your data which would make your hash-ing algorithm optimal.

(3.14) [QuickSort] Investigate different key selection strategies for the quick-sort algorithm. Test out at least five different strategies and use large lists of random data as your performance benchmark. Compare each strategy and rate the strategies in terms of their performance.

(3.15) [Simulated Annealing] Modify Code List 3.36 to use simulated anneal-ing to pack a number of rectangles into a rectangle with smallest area. Sup-port the option to pack rectangles into a square with smallest area.

4 Algorithms for Computer Arithmetic

4.1 2's Complement Addition

This section presents the principles of addition, multiplication and division for fixed point 2's complement numbers. Examples are provided for a selection of important fixed point algorithms.

Two's complement addition generates the sum, S, for the addition of two n-bit numbers A and B with

$$A = a_{n-1}...a_0$$

$$B = b_{n-1}...b_0$$

$$S = s_{n-1}...s_0$$

A C++ program simulating 8-bit two's complement addition is shown in Code List 4.1. The output of the program is shown in Code List 4.2

Code List 4.1 2's Complement Addition

C++ Source Code
#include <iostream.h>
unsigned char a[] = {0xfa,0x13,0xc4,0xff,0x80};
unsigned char b[] = {0x06, 0xdf,0xa6,0xfe,0x80};
#define NINPUTS (sizeof(a)/sizeof(unsigned char))
void main()
{
unsigned char sum;

Code List 4.1 2's Complement Addition (continued)

C++ Source Code

```
int overflow, carry;
register int i;
unsigned char add(unsigned char augend,unsigned char addend,
                  int * ovflp,int *carryp);


int b2w(unsigned char n);
carry=0;
for(i=0;i<NINPUTS;i++)
{
cout << "a= " << b2w(a[i]) << " b= " << b2w(b[i]) << endl;
sum = add(a[i], b[i], &overflow, &carry);
cout << "sum = " << b2w(sum) << " overflow= " << overflow <<
        " carry= " << carry << endl << endl;
}
}
unsigned char add(unsigned char augend,unsigned char addend,
                  int * ovflp,int *carryp)
{
unsigned int sum;
unsigned char rtn;
sum = augend+addend;
rtn = sum;
if(carryp!=0) *carryp=(sum&0x100)>>8;
if(ovflp!=0) *ovflp=(((augend&0x80)==(addend&0x80))
                        &&((sum&0x80)!=(augend&0x80)));
return rtn;
}
int b2w(unsigned char n)
{
return (((n&0x80)==0)?(n&0xff):(n|(-1<<8)));
}
```

Code List 4.2 Output of Program in Code List 4.1

C++ Output
a= −6 b= 6
sum = 0 overflow= 0 carry= 1
a= 19 b= −33
sum = −14 overflow= 0 carry= 0
a= −60 b= −90
sum = 106 overflow= 1 carry= 1
a= −1 b= −2
sum = −3 overflow= 0 carry= 1
a= −128 b= −128
sum = 0 overflow= 1 carry= 1

The programs do not check for overflow but simply simulate the additon as performed by hardware.

4.1.1 Full and Half Adder

In order to develop some fast algorithms for multiplication and addition it is necessary to analyze the process of addition and multiplication at the bit level. Full and half adders are bit-level building blocks that are used to perform addition.

A half adder is a module which inputs two signals, a_i and b_i, and generates a sum, s_i, and a carry-out c_i. A half adder does not support a carry-in. The outputs are as in Table 4.1.

TABLE 4.1 Half Adder Truth Table

Input		Output	
a_i	b_i	s_i	c_i
0	0	0	0
0	1	1	0

TABLE 4.1 Half Adder Truth Table (continued)

Input		Output	
a_i	b_i	s_i	c_i
1	0	1	0
1	1	0	1

A full adder has a carry-in input, c_i. A full adder is shown in Table 4.2.

TABLE 4.2 Full Adder Truth Table

Input			Output	
a_i	b_i	c_{i-1}	s_i	c_i
0	0	0	0	0
0	0	1	1	0
0	1	0	1	0
0	1	1	0	1
1	0	0	1	0
1	0	1	0	1
1	1	0	0	1
1	1	1	1	1

The full adder and half adder modules are shown in Figure 4.1. The boolean equation for the output of the full adder is

$$s_i = \overline{a_i}\overline{b_i}c_{i-1} + \overline{a_i}b_i\overline{c_{i-1}} + a_i\overline{b_i}\overline{c_{i-1}} + a_ib_ic_{i-1} \tag{4.1}$$

$$c_i = \overline{a_i}b_ic_{i-1} + a_i\overline{b_i}c_{i-1} + a_ib_i\overline{c_{i-1}} + a_ib_ic_{i-1} \tag{4.2}$$

The boolean equation for the output of the half adder is

$$s_i = \overline{a_i}b_i + a_i\overline{b_i} = a_i \oplus b_i \tag{4.3}$$

where $\oplus$ denotes the exclusive-or operation.

$$c_i = a_ib_i \tag{4.4}$$

The output delay of each module can be expressed in terms of the gate delay, Δ, of the technology used to implement the boolean expression. The sum, s_i,

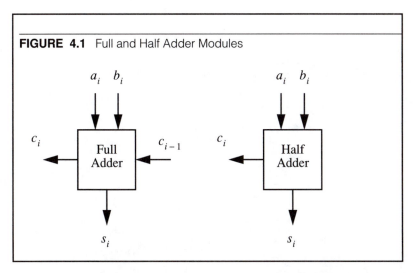

FIGURE 4.1 Full and Half Adder Modules

for the full adder can be implemented as in Eq. 4.1 using four 3-input NAND gates in parallel followed by a 4-input NAND gate. The gate delay of a k-input NAND gate is Δ so the sum is calculated in 2Δ. This is illustrated in Figure 4.2. For the half-adder the sum is calculated within 1Δ and the carry is generated within 1Δ. The Output Delay for the Half Adder is shown in Figure 4.2.

4.1.2 Ripple Carry Addition

2's complement addition of n-bit numbers can be performed by cascading Full Adder modules and a Half Adder module together as shown with a 4-bit example in Figure 4.3. The carry-out of each module is passed to the carry-in of the subsequent module. The output delay for an n-bit ripple-carry adder using a Half Adder module in the first stage is

$$\text{Output Delay} = (2n - 1)\Delta$$

For many applications this delay is unacceptable and can be improved dramatically.

A C++ program to perform ripple carry addition is shown in Code List 4.3. The output of the program is shown in Code List 4.4. The program demonstrates the addition of $1 + (-1)$. As can be seen in the output the carry ripples through the result at each simulation until it has passed over N bits.

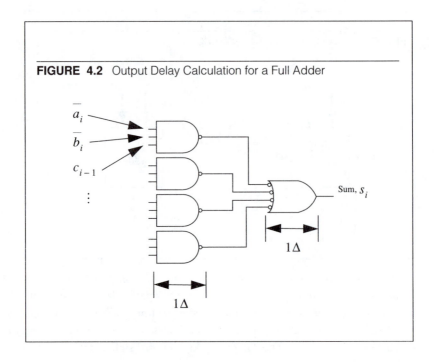

FIGURE 4.2   Output Delay Calculation for a Full Adder

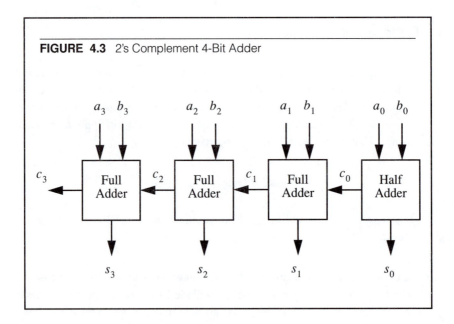

FIGURE 4.3   2's Complement 4-Bit Adder

FIGURE 4.4 Output Delay Calculation for a Half Adder

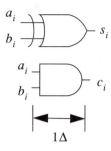

$$1\Delta$$

Code List 4.3 Ripple Carry Addition

C++ Source Code

```
// This program simulates Ripple Carry Addition

#include <iostream.h>
#define N 16
class DATA
        {
        public:
        char a,b, carry_in, sum, carry_out;
        };

class FADDER
        {
         public:
        DATA d_old, d_new;
        FADDER(){ d_old.a=d_old.b=d_old.carry_in
                =d_old.sum=d_old.carry_out=0;}
        void add();
        void update_data() {d_old=d_new;}
        };

void FADDER::add()
```

Code List 4.3 Ripple Carry Addition (continued)

C++ Source Code

```cpp
        {
        d_new.sum=d_old.a^d_old.b^d_old.carry_in;
        d_new.carry_out=d_old.a&d_old.b | d_old.a&d_old.carry_in
                        | d_old.b&d_old.carry_in;
        d_new.a=d_old.a;
        d_new.b=d_old.b;
        d_new.carry_in=d_old.carry_in;
        }


void set_carry_in(FADDER * f)
        {
        int i;
        for(i=1;i<N;i++)
                f[i].d_old.carry_in=f[i-1].d_old.carry_out;
        }

void set_data(int A, int B, FADDER * f)
        {
        int i;
        unsigned int mask=0x1;
        for(i=0;i<N;i++)
                {
                f[i].d_old.a=A&mask;
                f[i].d_old.b=B&mask;
                A>>=1;
                 B>>=1;
                }
        }

void full_add(FADDER *f)
        {
```

Code List 4.3 Ripple Carry Addition (continued)

C++ Source Code

```
            int i;
            for(i=0;i<N;i++) f[i].add();
            }


void update(FADDER *f)
            {
            int i;
            for(i=0;i<N;i++) f[i].update_data();
    }


void print_data(FADDER *f)
            {
            int i;
            cout << "A = ";
            for(i=N-1;i>=0;i--) cout << (f[i].d_old.a? "1":"0");
            cout << " B = ";
            for(i=N-1;i>=0;i--) cout << (f[i].d_old.b? "1":"0");
            cout << " SUM = ";
            for(i=N-1;i>=0;i--) cout << (f[i].d_old.sum? "1":"0");
            cout << endl;
    }



void main()
            {
            FADDER f[N];
            set_data(1,-1,f);
            print_data(f);
            int i;
            for(i=0;i<N;i++)
            {
                    set_carry_in(f);
```

Code List 4.3 Ripple Carry Addition (continued)

C++ Source Code
full_add(f);

```cpp
                full_add(f);
                update(f);
                print_data(f);
        }
    }
```

Code List 4.4 Output of Program in Code List 4.3

C++ Output
A = 0000000000000001 B = 1111111111111111 SUM = 0000000000000000
A = 0000000000000001 B = 1111111111111111 SUM = 1111111111111110
A = 0000000000000001 B = 1111111111111111 SUM = 1111111111111100
A = 0000000000000001 B = 1111111111111111 SUM = 1111111111111000
A = 0000000000000001 B = 1111111111111111 SUM = 1111111111110000
A = 0000000000000001 B = 1111111111111111 SUM = 1111111111100000
A = 0000000000000001 B = 1111111111111111 SUM = 1111111111000000
A = 0000000000000001 B = 1111111111111111 SUM = 1111111110000000
A = 0000000000000001 B = 1111111111111111 SUM = 1111111100000000
A = 0000000000000001 B = 1111111111111111 SUM = 1111111000000000
A = 0000000000000001 B = 1111111111111111 SUM = 1111110000000000
A = 0000000000000001 B = 1111111111111111 SUM = 1111100000000000
A = 0000000000000001 B = 1111111111111111 SUM = 1111000000000000
A = 0000000000000001 B = 1111111111111111 SUM = 1110000000000000
A = 0000000000000001 B = 1111111111111111 SUM = 1100000000000000
A = 0000000000000001 B = 1111111111111111 SUM = 1000000000000000
A = 0000000000000001 B = 1111111111111111 SUM = 0000000000000000

4.1.2.1 Overflow

The addition of two numbers may result in an overflow. There are four cases for the generation of overflow in 2's complement addition:

- Positive Number + Positive Number (result may be too large)
- Positive Number + Negative Number
- Negative Number + Positive Number
- Negative Number + Negative Number (result may be too negative)

Overflow is not possible when adding numbers with opposite signs. Overflow occurs if two operands are positive and the sum is negative or two operands are negative and the sum is positive. This results in the boolean expression

$$\text{Overflow} = a_{n-1} b_{n-1} \overline{s_{n-1}} + \overline{a_{n-1}} \, \overline{b_{n-1}} s_{n-1} \tag{4.5}$$

The calculation of overflow for ripple-carry addition can be simplified by analyzing the carry-in and carry-out to the final stage of the addition. This is demonstrated in Table 4.3. An overflow occurs when

TABLE 4.3 Carry Analysis for Overflow Detection

a_{n-1}	b_{n-1}	s_{n-1}	c_{n-1}	c_{n-2}	Overflow
0	0	0	0	0	0
0	0	1	0	1	1
1	1	0	1	0	1
1	1	1	1	1	0

$$c_{n-1} \neq c_{n-2} \tag{4.6}$$

which results in the boolean expression

$$\text{Overflow} = c_{n-1} \oplus c_{n-2} \tag{4.7}$$

4.1.3 Carry Lookahead Addition

In order to improve on the performance of the ripple-carry adder the carry-in to each stage is predicted in advance rather than waiting for the carry-in to propagate from the previous stages. The carry-out of each stage can be simplified from Eq. 4.2 to

$$c_i = a_i b_i + a_i c_{i-1} + b_i c_{i-1} \tag{4.8}$$

or

$$c_i = a_i b_i + (a_i + b_i) c_{i-1} \tag{4.9}$$

which is written as

$$c_i = g_i + p_i c_{i-1} \qquad (4.10)$$

with

$$g_i = a_i b_i$$

$$p_i = a_i + b_i$$

The interpretation of Eq. 4.10 is that at stage i a carry may be generated by the stage, $(g_i = 1)$, or a carry may be propagated from a previous stage, $(p_i = 1)$. When $g_i = 1$ stage i will always have a carry-out regardless of the carry-in. When $g_i = 0$ stage i will have a carry when the carry-in is 1 and $p_i = 1$, thus it is said to have propagated the carry. The time required to produce the generate, g_i, and the propagate, p_i, is 1Δ. For the a four-bit adder as in Figure 4.3 one has

$$c_0 = g_0 \qquad (4.11)$$

$$c_1 = g_1 + p_1 c_0 \qquad (4.12)$$

$$c_2 = g_2 + p_2 c_1 = g_2 + p_2 g_1 + p_2 p_1 g_0 \qquad (4.13)$$

$$c_3 = g_3 + p_3 c_2 = g_3 + p_3 g_2 + p_3 p_2 g_1 + p_3 p_2 p_1 g_0 \qquad (4.14)$$

The interpretation of Eq. 4.14 is that a carry-out will occur from stage 3 of the 4-bit adder if it is

- generated in stage 3
- generated in stage 2 and propagated through stage 3
- generated in stage 1 and propagated through stage 2 and stage 3
- generated in stage 0 and propagated through stage 1 and stage 2 and stage 3

The carry of the final stage, c_3, can be generated in 2Δ as shown in Figure 4.5. Similarly, the other carries can be calculated in 2Δ or less.

Once the carries are known the sum can be generated within 2Δ. Thus for the four bit adder the sum can be generated in a total of 5Δ with

- 1Δ to calculate the generates and propagates
- 2Δ to calculate the carries
- 2Δ to calculate the sums

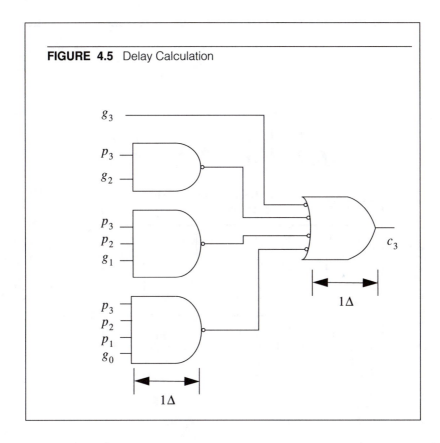

FIGURE 4.5 Delay Calculation

Using ripple-carry the four bit adder would require 7Δ to form the result. With the CLA adder the carries are thus generated by separate hardware. As is common, speed is thus achieved at the cost of additional hardware. The 4-bit CLA adder module is shown in Figure 4.3.

The CLA approach can be extended to n-bits yielding the following equation for the carry bits

$$c_i = \sum_{j=0}^{i} \left(\prod_{k=j+1}^{i} p_k \right) g_j \qquad (4.15)$$

with the product term evaluating to one when the indices are inconsistent. The calculation of the carries in Eq. 4.15 can be accomplished in 2Δ once the generates and propagates are known; however, there is a hardware requirement to be met. For each carry of the stage the implementation in 2Δ requires that the

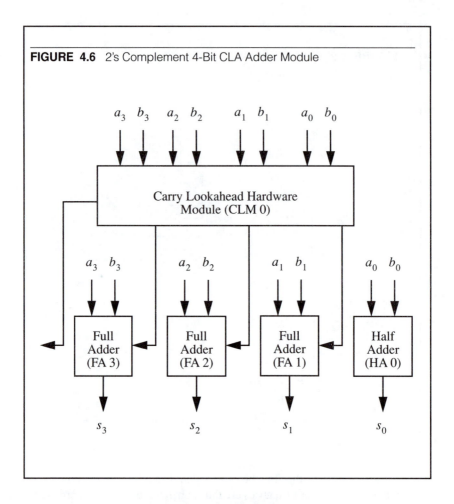

FIGURE 4.6 2's Complement 4-Bit CLA Adder Module

gates have a fan-in (number of inputs, to the gate) of $i + 1$. For an n-bit CLA adder realized in this manner a gate with a fan-in of n is required. This can be seen in Figure 4.5 where for a 4-bit CLA adder the carry inputs are calculated using a 4-input NAND gate. While this is practical for a 4-bit adder it is not practical for a 64-bit adder. As a result of this an inductive approach is needed to limit the fan-in requirements of the gates to implement the circuit. The timing of the 4-bit CLA adder module is shown in Figure 4.7.

When an inductive approach is taken the module shown in Figure 4.3 will need to input a carry in to the lowest stage. As a result the basic building block will be as shown in Figure 4.3. The module will be depicted as shown in Figure 4.8. The module serves as a basic building block for a 16-bit CLA adder as shown in Figure 4.10. For this case there are four groups of CLA-4

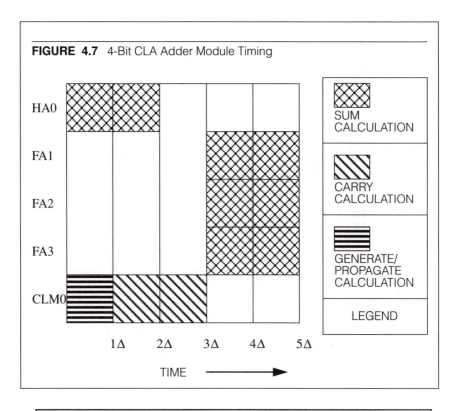

FIGURE 4.7 4-Bit CLA Adder Module Timing

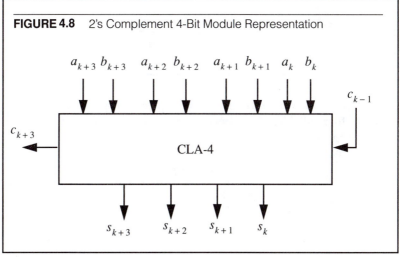

FIGURE 4.8 2's Complement 4-Bit Module Representation

building blocks. The carry lookahead hardware module CLM ($15 \rightarrow 0$) provides the carry input to each of the groups. This carry is predicted in an analo-

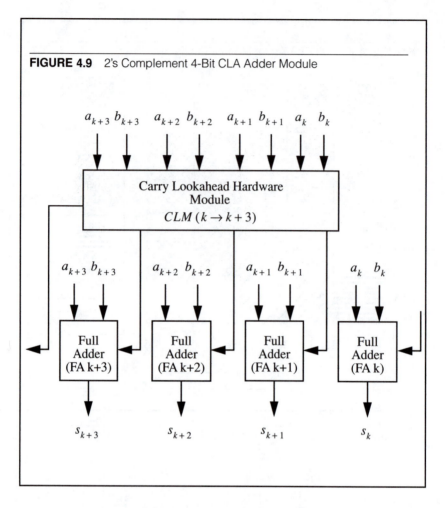

FIGURE 4.9   2's Complement 4-Bit CLA Adder Module

gous fashion to before. Group 0 will generate a carry if it is generated by one of the four individual full adders within the group. One can define group generate, gg_0, as

$$gg_0 = g_3 + p_3 g_2 + p_3 p_2 g_1 + p_3 p_2 p_1 g_0 \tag{4.16}$$

and group propagate, gp_0, as

$$gp_0 = p_3 p_2 p_1 p_0 \tag{4.17}$$

Similarly,

$$gg_1 = g_7 + p_7 g_6 + p_7 p_6 g_5 + p_7 p_6 p_5 g_4 \tag{4.18}$$

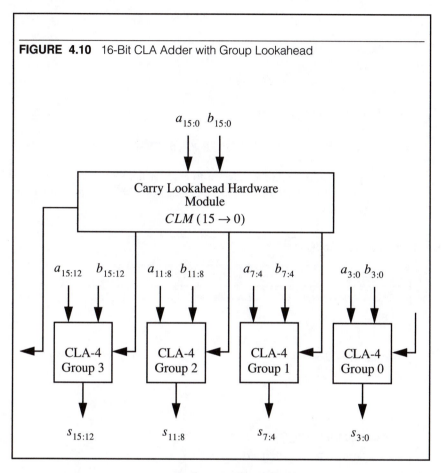

FIGURE 4.10 16-Bit CLA Adder with Group Lookahead

$$gp_1 = p_7p_6p_5p_4 \tag{4.19}$$

$$gg_2 = g_{11} + p_{11}g_{10} + p_{11}p_{10}g_9 + p_{11}p_{10}p_9g_8 \tag{4.20}$$

$$gg_3 = g_{15} + p_{15}g_{14} + p_{15}p_{14}g_{13} + p_{15}p_{14}p_{13}g_{12} \tag{4.21}$$

$$gp_3 = p_{15}p_{14}p_{13}p_{12} \tag{4.22}$$

From these equations one can derive the group carries as gc_0, the carry out of group 0,

$$gc_0 = gg_0, \tag{4.23}$$

gc_1, the carry out of group 1,

$$gc_1 = gg_1 + gp_1 gg_0, \qquad \text{(4.24)}$$

gc_2, the carry out of group 2,

$$gc_2 = gg_2 + gp_2 gg_1 + gp_2 gp_1 gg_0 \qquad \text{(4.25)}$$

gc_3, the carry out of group 3,

$$gc_3 = gg_3 + gp_3 gg_2 + gp_3 gp_2 gg_1 + gp_3 gp_2 gp_1 gp_0 \qquad \text{(4.26)}$$

The group carries become the carry-in to each of the CLA-4 modules. Each CLA-4 module can calculate the individual carries within 2Δ after the group carries are known.

Code List 4.5 CLA Addition

C++ Source

```
/*This code simulates a 64-bit CLA adder with 4 Sections, 16 Groups
64 Full Adders */


#include <stdio.h>
#include <iostream.h>


#define ADDER_SIZE 64
#define NUMBER_OF_GROUPS 16
#define NUMBER_OF_SECTIONS 4
#define SECTION_SIZE NUMBER_OF_GROUPS/NUMBER_OF_SECTIONS
#define GROUP_SIZE ADDER_SIZE/NUMBER_OF_GROUPS
#define ZERO 0
#define ONE 1


void get_data();
void print_signal();
void calc_gen_prop();
void calc_group_gen_prop();
void calc_section_carries();
void main();


typedef int SIGNAL;
```

Code List 4.5 CLA Addition (continued)

C++ Source

```cpp
void get_data(SIGNAL *a,SIGNAL *b)
{
unsigned long a_high, a_low, b_high, b_low;
unsigned int mask;
int i;
a_high=0xf0f0f0f0;
a_low=0xf0f0f0f0;
b_high=0x00000000;
b_low=0xffffffff;
for(i=0;i<64;i++)
        {
        a[i]=ZERO;
        b[i]=ZERO;
        }
mask=1;
for(i=0;i<32;i++)
        {
        if(mask&a_low) a[i]=ONE;
        if(mask&b_low) b[i]=ONE;
        mask=mask<<1;
        }
mask=1;
for(i=32;i<64;i++)
        {
        if(mask&a_high) a[i]=ONE;
        if(mask&b_high) b[i]=ONE;
        mask=mask<<1;
        }
}
void print_signal(SIGNAL * a,int len)
{
int i;
```

Code List 4.5 CLA Addition (continued)

C++ Source

```
for(i=len–1;i>=0;i– –) cout << (a[i]==ZERO)?"0":"1";

}


void calc_gen_prop(SIGNAL *a,SIGNAL *b,
        SIGNAL *generate,SIGNAL *propagate)
{
int i;
for(i=0;i<ADDER_SIZE;i++)
        {
        generate[i]=a[i]&&b[i];
        propagate[i]=a[i]||b[i];
        }
}
void calc_group_gen_prop(SIGNAL * generate, SIGNAL * propagate,
        SIGNAL * group_generate, SIGNAL * group_propagate)
{
int i,j,k;
SIGNAL partial_product, sum;
for(i=0;i<NUMBER_OF_GROUPS;i++)
        {
        group_generate[i]=ZERO;
        for(j=0;j<GROUP_SIZE;j++)
                {
                partial_product=generate[GROUP_SIZE*i+j];
                for(k=1; k< GROUP_SIZE–j; k++)
                        {
                        partial_product=partial_product
                                &&propagate[GROUP_SIZE*i+j+k];
                        }
                group_generate[i] = group_generate[i]||partial_product;
                }
```

Code List 4.5 CLA Addition (continued)

```
C++ Source
            }
for(i=0;i<NUMBER_OF_GROUPS;i++)
        {
        group_propagate[i]=ONE;
        for(j=0;j<GROUP_SIZE;j++)
                group_propagate[i]=group_propagate
                        &&propagate[GROUP_SIZE*i+j];
        }
}


void calc_section_gen_prop(SIGNAL * group_gen, SIGNAL * group_prop,
                SIGNAL *section_gen, SIGNAL *section_prop)
{
int i,j,k;
SIGNAL partial_product;

for(i=0;i<NUMBER_OF_SECTIONS;i++)
        {
        section_gen[i]=ZERO;
        for(j=0;j<SECTION_SIZE;j++)
                {
                partial_product=group_gen[SECTION_SIZE*i+j];
                for(k=1; k< SECTION_SIZE–j; k++)
                        {
                        partial_product=
                                partial_product&&
                                group_prop[SECTION_SIZE*i+j+k];
                        }
                section_gen[i] = section_gen[i]||partial_product;
                }
        }
for(i=0;i<NUMBER_OF_SECTIONS;i++)
```

Code List 4.5 CLA Addition (continued)

```cpp
        {
        section_prop[i]=ONE;
        for(j=0;j<SECTION_SIZE;j++)
                    section_prop[i]= section_prop[i]
                            &&group_prop[SECTION_SIZE*i+j];


        }
}


void calc_section_carries(SIGNAL * sec_carry,SIGNAL * sec_gen,
                        SIGNAL *sec_prop)
{
int i,j,k;
SIGNAL partial_product;
        sec_carry[0]=sec_gen[0];
for(i=1;i<NUMBER_OF_SECTIONS;i++)
        {
        sec_carry[i]=sec_gen[i]||(sec_prop[i]&&sec_carry[i–1]);
        }
}


void calc_group_carries(SIGNAL * group_carries, SIGNAL * group_gen,
                        SIGNAL * group_prop,SIGNAL * sec_carries)
{
int i, j;
for(i=0;i<NUMBER_OF_SECTIONS;i++)
        {

        i?(group_carries[i*SECTION_SIZE]=group_gen[i*SECTION_SIZE]||
            group_prop[i*SECTION_SIZE]&&sec_carries[i–1]):
            (group_carries[0]=group_gen[0]);
```

Code List 4.5 CLA Addition(continued)

C++ Source

```cpp
            for(j=1;j<SECTION_SIZE;j++)
                {
                group_carries[i*SECTION_SIZE+j]=
                        group_gen[i*SECTION_SIZE+j]||
                        group_prop[i*SECTION_SIZE+j]
                        &&group_carries[i*SECTION_SIZE+j-1];


                }
            }


}

void calc_carries(SIGNAL * carry,SIGNAL * gen,
                        SIGNAL *prop,SIGNAL *group_carry)
{
int i,j;
        for(i=0;i<NUMBER_OF_GROUPS;i++)
        {
        i?(carry[i*GROUP_SIZE]=gen[i*GROUP_SIZE]||
                prop[i*GROUP_SIZE]
                &&group_carry[i-1]):(carry[0]=gen[0]);

        for(j=1;j<GROUP_SIZE;j++)
                {

                carry[i*GROUP_SIZE+j]=
                        gen[i*GROUP_SIZE+j]||
                        prop[i*GROUP_SIZE+j]
                        &&carry[i*GROUP_SIZE+j-1];


                }
            }
```

Code List 4.5 CLA Addition (continued)

C++ Source

```cpp
}

void adder(SIGNAL *sum, SIGNAL a,SIGNAL b,SIGNAL c)
{
(a^b^c)?(*sum=ONE):(*sum=ZERO);
}
void calc_sum(SIGNAL * SUM,SIGNAL *A,SIGNAL *B,SIGNAL *CARRY)
{
int i;
adder(SUM,A[0],B[0],0);
for(i=1;i<ADDER_SIZE;i++)
adder(SUM+i,A[i],B[i],CARRY[i–1]);
}

void print_results(SIGNAL *A,SIGNAL *B,SIGNAL *GENERATE,
        SIGNAL *PROPAGATE,SIGNAL *GROUP_GENERATE,
        SIGNAL *GROUP_PROPAGATE,SIGNAL *SECTION_GENERATE,
        SIGNAL *SECTION_PROPAGATE,SIGNAL *SECTION_CARRY,
        SIGNAL *GROUP_CARRY,SIGNAL *CARRY,SIGNAL *SUM)
{
cout << "A = ";
print_signal(A,64);
cout << endl << "B = ";
print_signal(B,64);
cout << endl << "SUM = ";
print_signal(SUM,64);
cout << endl << "CARRY = ";
print_signal(CARRY,64);
cout << endl << "GENERATE = ";
print_signal(GENERATE,64);
cout << endl << "PROPAGATE = ";
print_signal(PROPAGATE,64);
```

Code List 4.5 CLA Addition (continued)

C++ Source

```
cout << endl << "GROUP_GENERATE = ";
print_signal(GROUP_GENERATE,16);
cout << endl << "GROUP_PROPAGATE = ";
print_signal(GROUP_PROPAGATE,16);
cout << endl << "SECTION_GENERATE = ";
print_signal(SECTION_GENERATE,4);
cout << endl << "SECTION_PROPAGATE = ";
print_signal(SECTION_PROPAGATE,4);
cout << endl << "SECTION_CARRY = ";
print_signal(SECTION_CARRY,4);
cout << endl << "GROUP_CARRY = ";
print_signal(GROUP_CARRY,16);
cout << endl;
}

void main()
{

/* declare data*/
SIGNAL A[ADDER_SIZE], B[ADDER_SIZE];
SIGNAL GENERATE[ADDER_SIZE], PROPAGATE[ADDER_SIZE];
SIGNAL GROUP_GENERATE[NUMBER_OF_GROUPS],
        GROUP_PROPAGATE[NUMBER_OF_GROUPS];
SIGNAL SECTION_GENERATE[NUMBER_OF_SECTIONS];
SIGNAL SECTION_PROPAGATE[NUMBER_OF_SECTIONS];
SIGNAL SECTION_CARRY[NUMBER_OF_SECTIONS];
SIGNAL GROUP_CARRY[NUMBER_OF_GROUPS];
SIGNAL CARRY[ADDER_SIZE];
SIGNAL SUM[ADDER_SIZE];



get_data(A,B);
```

Code List 4.5 CLA Addition (continued)

C++ Source

```
calc_gen_prop(A,B,GENERATE,PROPAGATE);


calc_group_gen_prop(GENERATE,PROPAGATE,
        GROUP_GENERATE,GROUP_PROPAGATE);


calc_section_gen_prop(GROUP_GENERATE, GROUP_PROPAGATE,
        SECTION_GENERATE, SECTION_PROPAGATE);


calc_section_carries(SECTION_CARRY,
        SECTION_GENERATE,SECTION_PROPAGATE);


calc_group_carries(GROUP_CARRY,GROUP_GENERATE,
        GROUP_PROPAGATE,SECTION_CARRY);


calc_carries(CARRY,GENERATE,PROPAGATE,GROUP_CARRY);


calc_sum(SUM,A,B,CARRY);


print_results(A,B,GENERATE,PROPAGATE,
                GROUP_GENERATE,GROUP_PROPAGATE,
                SECTION_GENERATE, SECTION_PROPAGATE,
                SECTION_CARRY,GROUP_CARRY,CARRY,SUM);
}
```

Code List 4.6 Output of Program in Code List 4.5

C++ Output
A =
11111111111111110000111100001111111111111111111111110000111100001111
B =
11111111111111111111111111111111111111111111111111110000000000000000

Code List 4.6 Output of Program in Code List 4.5 (continued)

C++ Output
SUM =
1111111111111110000111100001111111111111111111111100000111100010000
CARRY =
1111111111111111111111111111111111111111111111111111110000000000001111
GENERATE =
1111111111111111111111111111111111111111111111111111110000111100001111
PROPAGATE =
111111111111111000011110000111111111111111111111110000000000000000
GROUP_GENERATE = 1111111111110101
GROUP_PROPAGATE = 1111010111110000
SECTION_GENERATE = 1110
SECTION_PROPAGATE = 1110
SECTION_CARRY = 1110
GROUP_CARRY = 1111111111110001

4.2 A Simple Hardware Simulator in C++

This section starts the implementation of a simple hardware simulator in C++. The simulator will be used to simulate the hardware required to implement the algorithms in the previous sections.

A simple boolean logic simulator is shown in Code List 4.7. The output of the program is shown in Code List 4.8. The program simulates the interconnection of gates and is used to demonstrate the behavior of a clocked D flip-flop.

The program simulates the behavior of the circuit by calculating new values in the simulation in terms of the old values. The old values are then updated and the process is performed again. The process continues until the new and old values are identical or until a terminal count has been reached. For this program a terminal count of 50 is used but it is never reached in this example.

The circuit that is implemented is shown in Figure 4.11. The program allows each net to have one of three values: 0,1, or 2. The values are as follows:

- 0: Logical 0
- 1: Logical 1

- 2: Cannot be determined, printed out as x

All the values in the NET structure are initialized to the unknown state 2. As the inputs, clock, and data propagate through the circuit the values are changed as they become determined.

The behavior of each gate is modelled by its associated function within the program. The gates input one of the three states. The output is determined according to the logical function. This is illustrated in Table 4.4 for the 2-input NAND gate for all nine possibilities of the inputs.

TABLE 4.4   2-Input NAND behavior.

NAND behavior		
x	y	f(x,y)
0	0	1
0	1	1
0	x	1
1	0	1
1	1	0
1	x	x
x	0	1
x	1	x
x	x	x

The output data is shown in the timing diagram in Figure 4.13. As can be seen in the figure the circuit behaves as expected. The Q and QBAR outputs remain unknown until the first rising edge of the clock and at that point the output Q reflects the value of DATA at the clock edge. Only subsequent rising edges of the clock cause the outputs to change. It is important to note that this specific test does not demonstrate the validity of the device as a D flip-flop. In the absence of a theoretical proof a considerable amount of additional testing is necessary.

There is another interesting point about the simulation which can cause problems in circuit design. By looking at the last clock rise in Code List 4.8 one notes that QBAR makes a zero to one transition one gate delay quicker than Q making the corresponding one to zero transition. This is illustrated in Figure 4.12. As a result, it is important to let the data stabilize prior to its use.

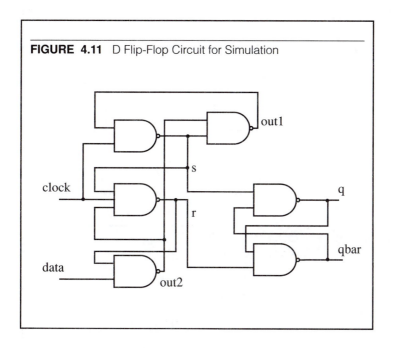

FIGURE 4.11   D Flip-Flop Circuit for Simulation

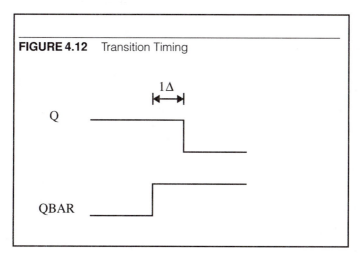

FIGURE 4.12   Transition Timing

4.3 2's Complement Multiplication

The goal of this section is to investigate algorithms for fast multiplication of two n-bit numbers to form a product. If two's complement notation is used

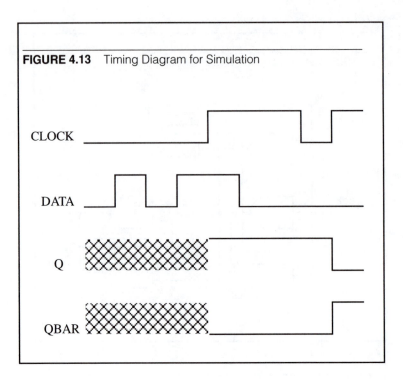

FIGURE 4.13   Timing Diagram for Simulation

Code List 4.7   Boolean Logic Simulator

C++ Simulator
/* This program implements a simple simulator for boolean logic */
#include <iostream.h>
class NET
{
public:
int new_value;
int old;
NET (int x=2, int y=2) { new_value=x; old=x;}
char print();
};
char NET::print()

```cpp
/* This program implements a simple simulator for boolean logic */
#include <iostream.h>

class NET
{
public:
int new_value;
int old;
NET (int x=2, int y=2) { new_value=x; old=x;}
char print();
};

char NET::print()
```

Code List 4.7 Boolean Logic Simulator (continued)

```
C++ Simulator
{
return (old? (old–1? 'x' : '1'):'0');
};


class NET clock, data, s, r, out1, out2, q, qbar;


int nand(int x, int y)
{
if((x==0)||(y==0)) return(1); else
        {
        if((x==1)&&(y==1))return(0);
        else return(2);
        }
}


int nand3(int x,int y,int z)
{
if((x==0)||(y==0)||(z==0)) return(1); else
        {
        if((x==1)&&(y==1)&&(z==1))return(0);
        else return(2);
        }
}
void update()
{
s.old=s.new_value;
r.old=r.new_value;
q.old=q.new_value;
qbar.old=qbar.new_value;
out1.old=out1.new_value;
out2.old=out2.new_value;
}
```

Code List 4.7 Boolean Logic Simulator (continued)

C++ Simulator

```
void print_result()
{
cout << "Clock " << clock.print() << " Data " << data.print()
        << " Q " << q.print() << " QBAR " << qbar.print() << endl;
}


void simulate(int cl,int da)
{
int stable=0,count=0;
clock.old=cl; data.old=da;
cout << "Clock = " << clock.print() << " Data = " << data.print() << endl;
while((!stable)&&(count<50))
        {
        stable=1;
        s.new_value=nand(out1.old,clock.old);
        r.new_value=nand3(s.old,clock.old,out2.old);
        out1.new_value=nand(s.old,out2.old);
        out2.new_value=nand(data.old,r.old);
        q.new_value=nand(qbar.old,s.old);
        qbar.new_value=nand(r.old,q.old);
        if(q.old != q.new_value) stable=0;
        if(qbar.old != qbar.new_value) stable=0;
        if(out1.old != out1.new_value) stable=0;
        if(out2.old != out2.new_value) stable=0;
        if(s.old != s.new_value) stable=0;
        if(r.old != r.new_value) stable=0;
        update();
        if((!stable)||(count==0)) print_result();
        count++;
        }
cout << "********************************" << endl;
```

Code List 4.7 Boolean Logic Simulator (continued)

```
C++ Simulator

}

void main()
{
simulate(0,0);
simulate(0,1);
simulate(0,0);
simulate(0,1);
simulate(1,1);
simulate(1,0);
simulate(1,0);
simulate(0,0);
simulate(1,0);
}
```

Code List 4.8 Output of Program in Code List 4.7

```
C++ Output

Clock = 0 Data = 0
Clock 0 Data 0 Q x QBAR x
Clock 0 Data 0 Q x QBAR x
*********************************
Clock = 0 Data = 1
Clock 0 Data 1 Q x QBAR x
Clock 0 Data 1 Q x QBAR x
*********************************
Clock = 0 Data = 0
Clock 0 Data 0 Q x QBAR x
Clock 0 Data 0 Q x QBAR x
*********************************
Clock = 0 Data = 1
Clock 0 Data 1 Q x QBAR x
```

Code List 4.8 Output of Program in Code List 4.7 (continued)

C++ Output
Clock 0 Data 1 Q x QBAR x
****
Clock = 1 Data = 1
Clock 1 Data 1 Q x QBAR x
Clock 1 Data 1 Q 1 QBAR x
Clock 1 Data 1 Q 1 QBAR 0
****
Clock = 1 Data = 0
Clock 1 Data 0 Q 1 QBAR 0
****
Clock = 1 Data = 0
Clock 1 Data 0 Q 1 QBAR 0
****
Clock = 0 Data = 0
Clock 0 Data 0 Q 1 QBAR 0
Clock 0 Data 0 Q 1 QBAR 0
****
Clock = 1 Data = 0
Clock 1 Data 0 Q 1 QBAR 0
Clock 1 Data 0 Q 1 QBAR 1
Clock 1 Data 0 Q 0 QBAR 1
****

then when multiplying two numbers, A and B,

$$A = a_{n-1}a_{n-2}...a_0 \tag{4.27}$$

$$B = b_{n-1}b_{n-2}...b_0 \tag{4.28}$$

In order to store the result one needs to calculate the number of bits required to represent the product in 2's complement form. By noting the range of 2's complement from Table 1.4 on page 11 one obtains that $2n$ bits are required in 2's complement form. The product is formed as

$$P = p_{2n-1}p_{2n-2}...p_0 \tag{4.29}$$

Since $2n$ bits are stored in the hardware for the product then overflow is not an issue.

4.3.1 Shift-Add Addition

The shift-add technique is the simple grade school technique for multiplication. In this scenario a partial product is formed by adding as appropriate repeated shifts of the multiplicand. The core statement in Code List 4.9 is

if(b&0x01) prod+=a; b=b>>1;a*=2;

This statement forms the product by repeatedly evaluating the lsb of the multiplier and if it is set by adding the shifted multiplicand. At each iteration the multiplier is shifted right to investigate the next bit and the multiplicand is shifted left.

Code List 4.9 Shift Add Technique

```
C++ Source

//This program demonstrates 2's complement multiplication using a
// shift-add technique
#include <stdio.h>
#include <iostream.h>
class operands {
// private data
private:
            int a,b,prod;
// public functions
public:
            void set_a(int x) { a=x;}
            void set_b(int x) { b=x;}
            void print_operands(void)
            {cout << "A= " << a << " B= " << b << endl;}
            void print_product(void)
                    {cout << "Product= " << prod << endl << endl;}
            void iterate(void)
                    {if(b&0x01) prod+=a; b=b>>1;a*=2;}
            void compute_product(void);
//declare constructor to initialize a,b,prod
```

Code List 4.9 Shift Add Technique (continued)

```
C++ Source
            operands(void){a=b=prod=0;};
        };


void operands::compute_product(void)
        {int i; prod=0; for(i=0;i<sizeof(int)*8;i++) iterate();}


int data[][2] = {{40,5}, {-20,57},{30,40},{-1,-4}};


void main()
{
        operands op; // here the private data is initialize to 0
        int i;
        for(i=0;i<sizeof(data)/sizeof(int)/2;i++)
        {
                op.set_a(data[i][0]);
                op.set_b(data[i][1]);
                op.print_operands();
                op.compute_product(); // This destroys the operands
                op.print_product();

        }
}
```

Code List 4.10 Output of Code List 4.9

```
C++ Output
A= 40 B= 5
Product= 200


A= -20 B= 57
Product= -1140


A= 30 B= 40
```

Code List 4.10 Output of Code List 4.9 (continued)

C++ Output
Product= 1200
A= –1 B= –4
Product= 4

4.3.2 Booth Algorithm

The Booth algorithm is a recoding technique which attempts to recode the multiplier to speedup the scenario where there are sequences of 1's. As an example consider the multiplication in base 10 of 9999*7. One can evaluate the result rather quickly by performing (10000–1)*7=69993. This can be done without the assistance of a computing device. The algorithm used is to recode the sequence of 9's and results in an operation which is much simpler. The technique can also be applied in binary. Instead of sequences of 9's however, one is interested in sequences of 1's.

The Booth algorithm is illustrated in Figure 4.14. In the figure the product is formed as the multiplication of A and B (A=14 and B=6). When the result is done A remains unchanged and the product is formed in P:B where the : operator indicates register concatenation. Register B no longer contains its initial value. This is written as

$$P:B \leftarrow A \times B \qquad (4.30)$$

The destruction of register B is common because it uses one less register to form the product. The Booth algorithm considers the lower order bit of register B in conjunction with the added bit which is initialized to zero. The bits determine the operation according to Table 4.6.

An example of booth recoding is illustrated in Table 4.5. In the worst case the Booth algorithm requires that n operations be performed to compute the product. This is illustrated in the last entry in Table 4.5. As a result the recoding operation for this operand has not simplified the problem. The average number of operations for a random operand by the algorithm is determined in Problem 4.10. Due to the average and worst-case complexity of the Booth algorithm a better solution is sought to find the product.

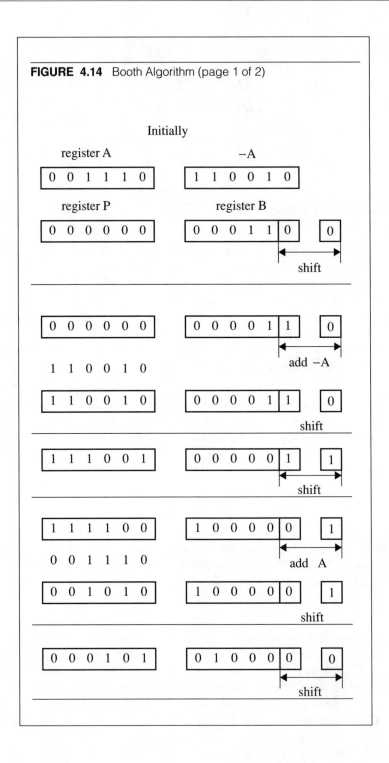

FIGURE 4.14 Booth Algorithm (page 1 of 2)

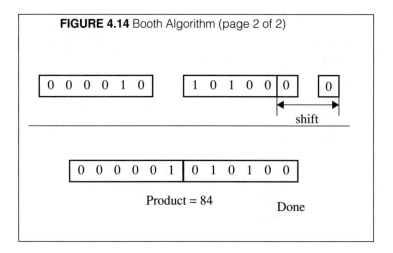

FIGURE 4.14 Booth Algorithm (page 2 of 2)

TABLE 4.5 Booth Recoding 8-Bit Example

Original Number								Booth Recode							
0	0	0	0	0	1	1	1	0	0	0	0	1	0	0	−1
0	0	0	0	1	1	0	0	0	0	0	1	0	−1	0	0
0	0	0	1	1	0	1	0	0	0	1	0	−1	1	−1	0
0	1	0	1	0	1	0	1	1	−1	1	−1	1	−1	1	−1

TABLE 4.6 Booth Recoding

Bit Pattern		Operation
0	0	product unchanged
0	1	product += a
1	0	product −= a
1	1	product unchanged

Code List 4.11 Booth Algorithm

C++ Source
//This program demonstrates 2's complement multiplication using a
// booth recoding technique

Code List 4.11 Booth Algorithm (continued)

C++ Source

```
#include <iostream.h>
class operands {
// private data
private:
          int a,b,prod;
// public functions
public:
          void set_a(int x) { a=x;}
          void set_b(int x) { b=x;}
          void print_operands(void)
          {cout << "A= " << a << " B= " << b << endl;}
          void print_product(void)
                    {cout << "Product= " << prod << endl << endl;}
          void iterate(void);
          void compute_product(void);
//declare constructor to initialize a,b,prod
          operands(void){a=b=prod=0;};
          };


void operands::iterate(void)
                    {
                              switch(b&0x3) {
                              case 1: prod+=a; break;
                              case 2: prod-=a; break;
                              default: break;
                              }

                    b=b>>1; a*=2;
                    }


void operands::compute_product(void)
          {
```

Code List 4.11 Booth Algorithm (continued)

```cpp
C++ Source
            int i; prod=0;
            if(b&0x1) prod-=a; a*=2;
            for(i=1;i<sizeof(int)*8;i++) iterate();
            }
int data[][2] = {{2,1}, {-20,57},{30,40},{-1,-4}};
void main()
{
            operands op; // here the private data is initialized to 0
            int i;
            for(i=0;i<sizeof(data)/sizeof(int)/2;i++)
            {
                        op.set_a(data[i][0]);
                        op.set_b(data[i][1]);
                        op.print_operands();
                        op.compute_product(); // This changes the operands a and b
                        op.print_product();

            }
}
```

Code List 4.12 Output of Program in Code List 4.11

```
C++ Program Output
A= 2 B= 1
Product= 2


A= -20 B= 57
Product= -1140


A= 30 B= 40
Product= 1200
```

Code List 4.12 Output of Program in Code List 4.11 (continued)

C++ Program Output
A= –1 B= –4
Product= 4

4.3.3 Bit-Pair Recoding

The Bit-Pair recoding technique is a technique which recodes the bits by considering three bits at a time. This technique will require $n/2$ additions or subtractions to compute the product. The recoding is illustrated in Table 4.7. The bits after recoding are looked at two at a time and the respective operations are performed. The higher order bit is weighted twice as much as the lower order bit. The C++ program to perform bit-pair recoding is illustrated in Code List 4.13. The output is shown in Code List 4.14.

The bit pair recoding algorithm is shown in Figure 4.14. The algorithm is analogous to the Booth recoding except that it investigates three bits at a time while the Booth algorithm looks at two bits at a time. The bit-pair recoding algorithm needs to have access to A, $-A$, 2A, and $-2A$ and as a result needs another additional 1-bit register to the left of P which is initialized to zero.

TABLE 4.7 Bit-Pair Recoding

Bit Pattern			Operation
0	0	0	no operation
0	0	1	$1 \times a$ $prod = prod + a;$
0	1	0	$2 \times a - a$ $prod = prod + a$
0	1	1	$2 \times a$ $prod = prod + 2a$
1	0	0	$-2 \times a$ $prod = prod - 2a$
1	0	1	$-2 \times a + a$ $prod = prod - a$
1	1	0	$-1 \times a$ $prod = prod - a$
1	1	1	no operation

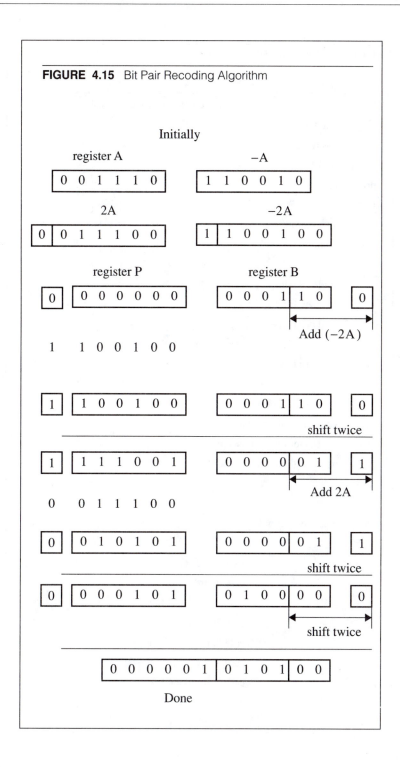

FIGURE 4.15 Bit Pair Recoding Algorithm

Code List 4.13 Bit-Pair Recoding Program

C++ Source

```cpp
//This program demonstrates 2's complement multiplication using a
// bit pair recoding technique
#include <iostream.h>
class operands {
// private data
private:
        int a,b,prod;
// public functions
public:
        void set_a(int x) { a=x;}
        void set_b(int x) { b=x;}
        void print_operands(void)
                {cout << "A= " << a << " B= " << b << endl;}
        void print_product(void)
                {cout << "Product = " << prod << endl << endl;}
        void iterate(void);
        void compute_product(void);
//declare constructor to initialize a,b,prod
        operands(void){a=b=prod=0;};


        };
void operands::iterate(void)
                {
                        switch(b&0x7) {
                        case 0: break;
                        case 1: prod+=a; break;
                        case 2: prod+=a; break;
                        case 3: prod+=2*a; break;
                        case 4: prod-=2*a; break;
                        case 5: prod-=a; break;
                        case 6: prod-=a; break;
                        case 7: break;
```

Code List 4.13 Bit-Pair Recoding Program (continued)

C++ Source

```
                              default: break;
                          }

                  b=b>>2; a*=4;
                  }
void operands::compute_product(void)
          {
          int i; prod=0;
// Take care of the first case which is special
          switch(b&0x3)
                          {
                  case 0: break;
                  case 1: prod+=a; break;
                  case 2: prod–=2*a; break;
                  case 3: break;
                   default: break;
                          }

          a*=4; b=b>>1;
          for(i=1;i<sizeof(int)*4;i++) iterate();
          }


int data[][2] = {{2,1}, {–20,57},{30,40},{–1,–4},{178,–178}};


void main()
{
          operands op; // here the private data is initialized to 0
          int i;
          for(i=0;i<sizeof(data)/sizeof(int)/2;i++)
          {
                  op.set_a(data[i][0]);
                  op.set_b(data[i][1]);
```

Code List 4.13 Bit-Pair Recoding Program (continued)

C++ Source

op.print_operands();
 	
op.compute_product(); // This changes the operands a and b
 	
op.print_product();
 	
    }
 	
}

Code List 4.14 Output of Program in Code List 4.13

C++ Output
A= 2 B= 1
Product = 2
A= –20 B= 57
Product = –1140
A= 30 B= 40
Product = 1200
A= –1 B= –4
Product = 4
A= 178 B= –178
Product = –31684

4.4   Fixed Point Division

This section presents algorithms for fixed point division. For fixed point division a $2n$ bit number, the dividend, is divided by an n bit number, the divisor, to yield an n bit quotient and an n bit remainder. Overflow can occur in the division process (see Problem 4.7).

4.4.1 Restoring Division

Restoring division is similar to the process of grade school addition. After aligning the bits appropriately the pseudocode is shown in Table 4.8.

TABLE 4.8 Division PsedudoCode

If $divisor < dividend$
$\quad${
$\quad\quad dividend = dividend - divisor$
$\quad\quad$place a 1 in quotient field
$\quad\quad$shift $dividend$ over
$\quad\quad$}
$\quad$else
$\quad\quad${
$\quad\quad$place a 0 in $quotient$
$\quad\quad$shift $dividend$ over
$\quad\quad$}

The pseudocode in Table 4.8 is repeated until the desired precision is reached. At which point the final dividend becomes the remainder. When this simple algorithm is executed on a computer in order for it to test whether $divisor < dividend$ it performs the subtraction

$$dividend = dividend - divisor \tag{4.31}$$

If the result is nonnegative then it places a 1 in the $quotient$ field. If the result is less than zero then the subtraction should not have occurred so the computer performs

$$dividend = dividend + divisor \tag{4.32}$$

to *restore* the dividend to the correct result and places a zero in the $quotient$ field. The computer then shifts the dividend and proceeds. This results in the pseudocode in Table 4.9.

TABLE 4.9 Restoring Division PseudoCode

$dividend = dividend - divisor$
if $dividend \geq 0$
$\quad${

TABLE 4.9 Restoring Division PseudoCode (continued)

place a 1 in *quotient* field
 	
}
else
 	
{
$dividend = dividend + divisor$
place a 0 in the *quotient* field
 	
}
shift over *dividend*

Problem 4.3 develops a C++ program to simulate restoring division.

4.4.2 Nonrestoring Division

Nonrestoring division is a technique which avoids the need to restore on each formation of the quotient bit. In effect, the need to restore is delayed until the final quotient bit is formed. The algorithm avoids this by noting that if a subtraction occurred that should not have then the next step in the algorithm would be to restore, then shift, then subtract.

$$dividend' = dividend - divisor \tag{4.33}$$

$$dividend'' = 2 \times (dividend' + divisor) - divisor \tag{4.34}$$

so that

$$dividend'' = 2 \times dividend' + divisor \tag{4.35}$$

It can be seen that the (restore,shift,subtract) is equivalent to a (shift,add). This is used to avoid the restore operation and is thus called nonrestoring division. The computer does continuous shift-subtract operations until the result is negative at which point the next operation becomes a shift-add. If on the final cycle the result is negative the computer will add the divisor back to restore the dividend (which on the final cycle is the remainder).

The program to perform nonrestoring division is shown in Code List 4.15. The output of the program is shown in Code List 4.16. The program uses a similar register-saving technique to the Booth algorithm. the program performs the division of a $2n$ bit number by an n bit number

$$\frac{R:Q}{B} \tag{4.36}$$

At the termination of the program the remainder is in R and the quotient is in Q. The program illustrates the division of 37/14 which yields 2 with a remainder of 9.

The program demonstrates a number of features in C++. The program introduces a class called *number* which defines the operations for the data. The class includes data and functions:

- *number*: this is the constructor function for the class which is called when a variable of type *number* is created
- *get_value*: the get_value function is used to return bit number x of the number. This is used to access the private data of the class which is hidden from the user.
- *shift_left*: the shift_left function is used to perform a logical left shift on the data. This operation is used extensively in the nonrestoring division algorithm.
- *print_value*: the function print_value is used to print the number and accepts a character string to be printed before prior to the value.
- *ones_complement*: the ones_complement function performs the ones_complement which is used to calculate the negative of a *number* in the addition process.
- *operator>=*: this overloads the greater than or equal operator in the program. When comparing two objects of type *number* this function is called.
- *operator<*: this operator overloads the less than operator when comparing objects of type *number*.
- *operator+*: this operator overloads the plus operator when comparing objects of type *number*.
- *operator−*: this operator overloads the minus operator when comparing objects of type *number*.

The + operator is defined first and is used in subsequent definitions of other overloaded operators. The + operator performs a ripple-carry (see Section 4.1.2) addition of the two numbers passed and returns the result as a number.

Rather than calculate the algorithm for the − operator it uses the newly overloaded + operator to calculate the subtraction by noting that $x - y = x + (-y)$.

The >= operator uses the newly formed – operator to return the difference in x and y as a number and accesses the most significant bit (the sign) of it to see if the difference is less than zero. It returns a value according to the test.

The < operator performs in a similar fashion.

The *left_shift_add* function introduces a feature of C++ not present in C. The first parameter in the function argument list is declared as *number& B*. As a result B is passed to the function as a pointer and is automatically dereferenced on use. See Section 3.1 for a more detailed description of pointers in C++.

Code List 4.15 Nonrestoring Division

C++ Source Code

```
#include <iostream.h>
#define N 32


class number
{
private:
        char value[N];


public:
        number(long x=0);
        char get_value(int x) { return value[x];}
        void shift_left();
        void print_value(char * x);
        number ones_complement();
        friend int operator>=(number x,number y);
        friend int operator<(number x,number y);
        friend number operator+(number x, number y);
        friend number operator–(number x, number y);
};


number number::ones_complement()
{
int i;
```

Code List 4.15 Nonrestoring Division (continued)

```cpp
C++ Source Code
number x;
for(i=0;i<N;i++) if(value[i]==0) x.value[i]=1; else x.value[i]=0;
return x;
}


void number::number(long x)
{
int i;
unsigned long mask=0x1;
for(i=0;i<N;i++)
        {       value[i]=(x&mask?1:0);
                mask<<=1;
        };


}


void number::shift_left()
{
int i;
        for(i=1;i<N;i++) value[N-i]=value[N-1-i];
 value[0]=0.0;
}


void number::print_value(char * x)
{
int i;
cout << x;
for(i=N-1;i>=0;i--) { cout << (char) (value[i]+0x30); }
cout << endl;
}


number operator+(number x, number y)
```

Code List 4.15 Nonrestoring Division (continued)

```
C++ Source Code
{
int i;
int carry=0;
for(i=0;i<N;i++)
        {
        switch (x.value[i]*4+y.value[i]*2+carry)
        {
                case 3:
                case 5:
                case 6:
                case 7:
                        x.value[i]=x.value[i]^y.value[i]^carry;
                        carry=1;
                        break;

                default: x.value[i]=x.value[i]^y.value[i]^carry;
                        carry=0;
                        break;
                }
            }
return x;
}


number operator-(number x, number y)
{
return (x+y.ones_complement()+1);
}


int operator>=(number x, number y)
{
if ((x-y).get_value(N-1)==1) return 0; else return 1;
}
```

Code List 4.15 Nonrestoring Division (continued)

C++ Source Code

```cpp
int operator<(number x,number y)
{
if ((x–y).get_value(N–1)==1) return 1; else return 0;
}



void left_shift_add(number& B, number& R, number& Q)
{
        R.shift_left();
        R=R+Q.get_value(N–1);
        Q.shift_left();
        R=R+B;
        if(R>=0) Q=Q+1;
}

void left_shift_subtract(number& B, number& R, number& Q)
{
        R.shift_left();
        R=R+Q.get_value(N–1);
        Q.shift_left();
        R=R–B;
        if(R>=0) Q=Q+1;
}

void restore(number& B, number& R)
{
        R=R+B;
}

void main()
{
```

Code List 4.15 Nonrestoring Division (continued)

```
  C++ Source Code
        number B(14),R(0),Q(37);
        int j;

        B.print_value("B = ");
        R.print_value("R = ");
        Q.print_value("Q = ");

        for(j=0;j<N;j++)
                {
                if(R>=0) left_shift_subtract(B,R,Q); else
                left_shift_add(B,R,Q);
                }
        if(R<0) restore(B,R);

        cout << "Calculation Done" << endl;
        R.print_value("R = ");
        Q.print_value("Q = ");
}
```

Code List 4.16 Output of Program in Code List 4.15

```
  C++ Output
  B = 00000000000000000000000000001110
  R = 00000000000000000000000000000000
  Q = 00000000000000000000000000100101
  Calculation Done
  R = 00000000000000000000000000001001
  Q = 00000000000000000000000000000010
```

4.4.3 Shifting over 1's and 0's

If the divisor is normalized so that it begins with a 1 then the technique of the previous sections can be improved to skip over 1's and 0's. Shifting over 0's is simple to see. If 0.000010101 is divided by 0.10111 It is easy to see that the first four quotient bits are zero. So rather than performing the subtraction, the dividend is renormalized each time a string of zero's is encountered. Similarly, if after each subtraction the result is a string of 1's, then the 1's can be skipped over placing 1's in the quotient bit. This technique is derived in Problem 4.5.

4.4.4 Newton's Method

In Newton's method the quotient to be formed is the product $A(1/B)$. For this case, once $1/B$ is determined a single multiplication cycle generates the desired result. Newton's method yields the iteration

$$x_{i+1} = x_i - \frac{f(x_i)}{f'(x_i)} \tag{4.37}$$

which for the function

$$f(x) = \frac{1}{x} - B \tag{4.38}$$

gives

$$x_{i+1} = x_i(2 - Bx_i) \tag{4.39}$$

Under suitable well known conditions x_i will converge to the inverse. Hence using Newton's algorithm the process of division is achieved via addition and multiplication operations. The C++ source code illustrating this technique is shown in Code List 4.17. The output of the program is shown in Code List 4.18.

Code List 4.17 Floating Point Division

```
C++ Source Code

#include <iostream.h>

#include <math.h>

// This program simulates Newton's method to perform

// The division A/B


class data

        {

private:

        double value;
```

Code List 4.17 Floating Point Division (continued)

```
C++ Source Code
double iter;
        public:
        data(double x=1.0) { value = x; iter=1.0;}
        void print() { cout << "Iteration value is " << iter <<endl; }
        void iterate() { iter = iter*(2–iter*value); }
        double error() { return fabs(iter–1.0/value); };
        double inverse() { return 1.0/value; };
void simulate();
        };


void data::simulate()
        {
cout << "Calculating inverse for x= " << value << endl;
        while(error()>1.0e-5) { iterate(); print();}
        cout << endl
                << "True inverse is 1/x=" << inverse() << endl;
        cout << "Error is " << error() <<
                << endl <<
                ****************************" << endl;
        }


void main()
        {
        data x;
        x=.7;
        x.simulate();
        x=.75;
        x.simulate();
        x=0.5;
        x.simulate();
        x=1.0;
```

Code List 4.17 Floating Point Division (continued)

C++ Source Code
x.simulate();   }

Code List 4.18 Output of Program in Code List 4.17

C++ Output
Calculating inverse for x= 0.7
Iteration value is 1.3
Iteration value is 1.417
Iteration value is 1.428478
Iteration value is 1.428571
True inverse is 1/x=1.428571
Error is 6.149532e-09
****
Calculating inverse for x= 0.75
Iteration value is 1.25
Iteration value is 1.328125
Iteration value is 1.333313
Iteration value is 1.333333
True inverse is 1/x=1.333333
Error is 3.104409e-10
****
Calculating inverse for x= 0.5
Iteration value is 1.5
Iteration value is 1.875
Iteration value is 1.992187
Iteration value is 1.999969
Iteration value is 2
True inverse is 1/x=2

Code List 4.18 Output of Program in Code List 4.17 (continued)

C++ Output
Error is 4.656613e-10
****
Calculating inverse for x= 1
True inverse is 1/x=1
Error is 0
****

4.5 Residue Number System

4.5.1 Representation in the Residue Number System

The residue number systems is a system which uses an alternate way to represent numbers. For integers, in 2's complement notation, the representation for a number was

$$A \equiv a_{n-1}a_{n-2}\ldots a_0 \tag{4.40}$$

with a value of

$$A = \left(\sum_{k=0}^{n-2} a_k 2^k \right) - a_{n-1} 2^{n-1} \qquad a_k \in \{0, 1\} \tag{4.41}$$

For this case, a number A is represented with n binary bits. The value is relatively easy to calculate via Eq. 4.41. A natural problem occurred with this representation for the process of addition. When n is large the calculation of the carry-in to each stage is the dominating factor with regard to the performance of the addition operation as noted in Section 4.1.2. Using methodologies in number theory, an alternate representation can be used which reduces the problems of with regard to the carry-in calculation.

The residue number system uses a set of relatively prime numbers:

$$M = \{m_0, m_1, \ldots, m_{n-1}\} \tag{4.42}$$

and represents a number A with respect to these moduli by the n-tuple:

$$A \equiv (A \bmod m_0, A \bmod m_1, \ldots, A \bmod m_{n-1}) \tag{4.43}$$

$$A \equiv (a_0, a_1, \ldots, a_{n-1}) \tag{4.44}$$

Two numbers are relatively prime if their greatest common divisor is one. Using the standard notation with

$$(x, y) \tag{4.45}$$

to denote the greatest common divisor of x and y. The requirement on the set M is that each of the members be pairwise relatively prime:

$$(m_i, m_j) = 1 \qquad 0 \le i, j \le n - 1 \tag{4.46}$$

For example, a representation with the moduli

$$M = \{2, 3, 5, 7, 11\} \tag{4.47}$$

the number 12 is represented as

$$(0, 0, 2, 5, 1) = 12 \tag{4.48}$$

and 14 is represented as

$$0, 2, 4, 0, 3) = 14 \tag{4.49}$$

The addition of 12 and 14 can be accomplished by adding the vector representation and performing the modulus operation:

$$(0, 0, 2, 5, 1) + (0, 2, 4, 0, 3) = ((0+0) \bmod 2, (0+2) \bmod 3, \ldots)$$
$$= (0, 2, 1, 5, 4) \tag{4.50}$$

Notice that the result is the same obtained when representing 26 in the notation.

The Range of the Residue Number Systems

The residue number system can represent N distinct numbers with

$$N = \prod_{i=0}^{n-1} m_i \tag{4.51}$$

For example, the moduli in Eq. 4.47,

$$N = 2 \times 3 \times 5 \times 7 \times 11 = 2310 \tag{4.52}$$

The result stated in Eq. 4.51 is established in Problem 4.15.

4.5.2 Data Conversion — Calculating the Value of a Number

This section derives a method for calculating the value of a number given only its representation in terms of the moduli. It is necessary to introduce some quantities in number theory. The Euler totient function, $\varphi(n)$, is defined for a number, n, as the number of positive integers satisfying

$$(n, k) = 1 \qquad 1 \le k \le n \tag{4.53}$$

For example,

$$
\begin{aligned}
\varphi(1) &= 1 \\
\varphi(2) &= 1 \\
\varphi(3) &= 2
\end{aligned}
\tag{4.54}
$$

If n is a prime number then

$$\varphi(n) = n - 1 \tag{4.55}$$

defining the weights, w_i, as

$$w_i = \left(\frac{N}{m_i}\right)^{\varphi(m_i)} \tag{4.56}$$

The vector W as

$$W = (w_0, w_1, ..., w_{n-1}) \tag{4.57}$$

and a number A, as

$$A = (a_0, a_1, ..., a_{n-1}) \tag{4.58}$$

The value of A is given as

$$\text{value}(A) = (W \cdot A) \bmod N = \left(\sum_{i=0}^{n-1} W_i m_i\right) \bmod N \tag{4.59}$$

This result is established in Problem 4.17. Consider the example in Eq. 4.47. For this case:

$$w_0 = \frac{N}{m_0} = \frac{N}{2} = 1155 \qquad (4.60)$$

Similarly, W becomes

$$W = (1155, 1540, 1386, 330, 210) \qquad (4.61)$$

To calculate the number 26 from its representation in Eq. 4.50 one has

$$\text{Value } (A) = (1155, 1540, 1386, 330, 210) \cdot (0, 2, 1, 5, 4)$$
$$= (2 \cdot 1540 + 1386 + 5 \cdot 330 + 4 \cdot 210) \bmod 2310 \qquad (4.62)$$
$$= 6956 \bmod 2310 = 26$$

4.5.3 C++ Implementation

A program to simulate the Residue Number System is shown in Code List 4.19. The output of the program is shown in Code List 4.20.

In the program a class *data* is declared which has the following data and functions:

- *unsigned moduli[N]:* this data item is used to hold the representation of each of the moduli.
- *data*: this is the constructor function for *data* which is called any time a variable is declared.
- *set:* this function is used to set the data's value.
- *print:* this function is used to print out the *moduli* and the value by calling the *value* function.
- *value:* this function calculates the value of the number from its residue representation.
- *operator+*: the + operator has been overloaded to perform the required addition in the residue number system.
- *operator**: the * operator has been overloaded to perform multiplication in the residue number system.

This program is a natural example for the use of the overloading operators in C++. Since the addition of the two numbers in the residue systems consists of the respective additions of their moduli it is natural to replace this operator for addition.

The output supplies all the moduli and prints out the relatively prime numbers at the top. Notice that the print function takes in an optional char * to print out a small string. If the string is not supplied it defaults to an empty string.

Code List 4.19 Residue Number System

C++ Source

```cpp
// This program simulates addition and multiplication in the residue number system
#include <iostream.h>
#include <iomanip.h>
#include <stdlib.h>
#include <string.h>
unsigned long rprime[]={7,15,31,32};
unsigned long eul[]={6,8,30,16};
unsigned long weights[4];
#define N sizeof(rprime)/sizeof(long)
class data
{
unsigned int moduli[N];
public:
          data(unsigned long x=0);
          void set(unsigned long x=0);
          void print(char * x = "");
          unsigned long value();
          friend data operator+(data x,data y);
          friend data operator*(data x,data y);
};
// constructor function
data::data(unsigned long x)
{
int i;
for(i=0;i<N;i++) moduli[i]=x%rprime[i];
}
void data::set(unsigned long x)
{
int i;
for(i=0;i<N;i++) moduli[i]=x%rprime[i];
}
```

Code List 4.19 Residue Number System (continued)

C++ Source

```cpp
void data::print(char * x)
{
int i;
cout << setw(7) << x;
for(i=0;i<N;i++) cout << " " << setw(2) << moduli[i];
cout << " * " << value() << endl;
}
unsigned long data::value()
{
int i;
unsigned long x=0;
unsigned long prod=1;
for(i=0;i<N;i++) prod*=rprime[i];
for(i=0;i<N;i++) x=(x+moduli[i]*weights[i])%prod;
return x;
}
//overload addition operator
data operator+(data a, data b)
        {
        data c;
        int i;
        for(i=0;i<N;i++) c.moduli[i] = (a.moduli[i]+b.moduli[i])%rprime[i];
        return c;
        }
//overload multiplication operator
data operator*(data a, data b)
        {
        data c;
        int i;
        for(i=0;i<N;i++) c.moduli[i] = (a.moduli[i]*b.moduli[i])%rprime[i];
        return c;
        }
```

Code List 4.19 Residue Number System (continued)

C++ Source

```cpp
void header()
{
int i;
long prod=1;
cout << setiosflags(ios::left);
for(i=0;i<N;i++) prod*=rprime[i]; cout << "Range Handled  0 to "
          << prod-1 << endl << endl;
cout << setw(7) << "Comment";
for(i=0;i<N;i++) cout<< "  " << setw(2) <<  rprime[i];
cout << "  *  Value" << endl;
for(i=0;i<34;i++) cout << "*"; cout << endl;
// Caclulate weights
for(i=0;i<N;i++)
        {
        unsigned long k;
        k=prod/rprime[i];
        weights[i]=1;
        int j;
        for(j=0;j<eul[i];j++) weights[i]=(weights[i]*k)%prod;
        }
}
void main()
{
header();
data x(29),y(30);
x.print("x=29");
y.print("y=30");
x=x+y;
x.print("x=x+y");
x=x+2;
x.print("x=x+2");
x=x*3;
```

Code List 4.19 Residue Number System (continued)

C++ Source

x.print("x=x*3");
 	
// Let's look at the weights
 	
    {
    int i;
    char s[8], num[2];
    for(i=0;i<N;i++)   { strcpy(s,"weight"); x.set(weights[i]);
           x.print(strcat(s,itoa(i,num,10))); }
    }
 	
}

Code List 4.20 Output of Program in Code List 4.19

C++ Output
Range Handled   0 to 104159
Comment   7   15   31   32   *   Value
****
x=29   1   14   29   29   *   29
y=30   2   0   30   30   *   30
x=x+y   3   14   28   27   *   59
x=x+2   5   1   30   29   *   61
x=x*3   1   3   28   23   *   183
weight0   1   0   0   0   *   44640
weight1   0   1   0   0   *   97216
weight2   0   0   1   0   *   43680
weight3   0   0   0   1   *   22785

Code List 4.21   Euler Totient Function

C++ Source

```cpp
#include <iostream.h>
// This program determines the Euler totient function
unsigned long rprime[]={7,15,31,32};
#define N sizeof(rprime)/sizeof(long)
unsigned long gcd(unsigned long x, unsigned long y)
{
while(y!=0) {
        unsigned temp=y;
        y=x%y;
        x=temp;
        }
return x;
}
void main()
{
unsigned long i,j,value;
for(i=0;i<N;i++)
        {
        value=0;
        for(j=1;j<rprime[i];j++) if(gcd(j,rprime[i])==1) value++;
        cout << "The value for " << rprime[i] << " is " << value << endl;
        }
}
```

Code List 4.22   Output of Program in Code List 4.21

C++ Output
The value for 7 is 6
The value for 15 is 8
The value for 31 is 30
The value for 32 is 16

4.6 Problems

(4.1) Modify Code List 4.1 to simulate 16, 32, and 64-bit 2's complement addition. Add a procedure to detect for overflow and indicate via output when overflow has occurred.

(4.2) Modify Code List 4.5 to simulate a CLA adder with 3 sections each with 3 groups each with 8 1-bit adders.

(4.3) Write a C++ program to simulate restoring division. Your program should support n bit inputs. Use the overload operators to perform addition and subtraction of each of the inputs.

(4.4) Modify the Code List 4.13 to support n bit inputs. Use a similar register structure as the example in Figure 4.14.

(4.5) First by example, then by proof, demonstrate the technique of shifting over 1's and 0's in non-restoring division.

(4.6) Write a C++ program to simulate modify Code List 4.15 to shift over 1's and 0's.

(4.7) Derive the conditions for overflow in fixed point division.

(4.8) Add all the common logical functions to Code List 4.7.

(4.9) Rewrite Code List 4.7 to simulate a JK Flip-Flop.

(4.10) Calculate the average number of operations required in the Booth algorithm for 2's complement multiplication. How does this compare to the shift-add technique?

(4.11) Modify Code List 4.7 to simulate Carry Lookahead Addition at the gate level for an 8-bit module.

(4.12) [Moderately Difficult] Modify Code List 4.13 to output, to a PostScript file, the timing diagram for the circuit which is simulated. Make rational assumptions about the desired interface. Use the program to generate a PostScript file for the timing diagram in Figure 4.12.

(4.13) Graphically illustrate Newton's method described in Eq. 4.37.

(4.14) Theoretically demonstrate that the *gcd* function in Code List 4.21 does in fact return the greatest common divisor of the inputs x and y.

(4.15) [Uniqueness] Show that if a residue number system is defined with moduli

$$M = \{m_0, m_1, \ldots, m_{n-1}\}$$

and A and B are integers such that

$$0 \le A < N \qquad 0 \le B < N \qquad N = \prod m_i$$

and if

$$a_i = b_i \qquad 0 \le i < N$$

with

$$a_i = A \bmod m_i \qquad b_i = B \bmod m_i$$

then

$$A = B$$

(4.16) If m_i and m_j are integers satisfying

$$(m_i, m_j) = (m_i - 1)\,\delta_{ij} + 1 \qquad \begin{array}{l} 0 \le i \le m - 1 \\ 0 \le j \le m - 1 \end{array}$$

with

$$\delta_{ij} = \begin{cases} 1, & (i = j) \\ 0, & \text{otherwise} \end{cases}$$

and

$$N = \prod_{i=0}^{n-1} m_i$$

prove that if

$$w_i = \left(\frac{N}{m_i}\right)^{\varphi(m_i)}$$

then

$$w_i \bmod m_j = \delta_{ij}$$

(4.17) Prove that Eq. 4.59 is true.

Index